NUCLEAR ENVIRONMENTAL CHEMICAL ANALYSIS

ELLIS HORWOOD SERIES IN ANALYTICAL CHEMISTRY

Series Editors: Dr. R. A. CHALMERS and Dr. MARY MASSON, University of Aberdeen

Consultant Editor: Prof. J. N. MILLER, Loughborough University of Technology

S. Allenmark — **Chromatographic Enantioseparation — Methods and Applications**
G.E. Baiulescu & V.V. Coşofreţ — **Application of Ion Selective Membrane Electrodes in Organic Analysis**
G.E. Baiulescu, P. Dumitrescu & P.Gh. Zugravescu — **Sampling**
G.E. Baiulescu, C. Patroescu & R.A. Chalmers — **Education and Teaching in Analytical Chemistry**
G.I. Bekov & V.S. Letokhov — **Laser Resonant Photoionization Spectroscopy for Trace Analysis**
S. Bance — **Handbook of Practical Organic Microanalysis**
H. Barańska, A. Łabudzińska & J. Terpiński — **Laser Raman Spectrometry**
K. Beyermann — **Organic Trace Analysis**
O. Budevsky — **Foundations of Chemical Analysis**
J. Buffle — **Complexation Reactions in Aquatic Systems: An Analytical Approach**
D.T. Burns, A. Townshend & A.G. Catchpole — **Inorganic Reaction Chemistry Volume 1: Systematic Chemical Separation**
D.T. Burns, A. Townshend & A.H. Carter — **Inorganic Reaction Chemistry: Reactions of the Elements and their Compounds: Volume 2, Part A: Alkali Metals to Nitrogen and Volume 2, Part B: Osmium to Zirconium**
S. Caroli — **Improved Hollow Cathode Lamps for Atomic Spectroscopy**
J. Churáček — **New Trends in the Theory & Instrumentation of Selected Analytical Methods**
R. Czoch & A. Francik — **Instrumental Effects in Homodyne Electron Paramagnetic Resonance Spectrometers**
T.E. Edmonds — **Interfacing Analytical Instrumentation with Microcomputers**
J.K. Foreman & P.B. Stockwell — **Automatic Chemical Analysis**
Z. Galus — **Fundamentals of Electrochemical Analysis**
J. Gasparič & J. Churáček — **Laboratory Handbook of Paper and Thin Layer Chromatography**
S. Görög — **Steroid Analysis in the Pharmaceutical Industry**
T. S. Harrison — **Handbook of Analytical Control of Iron and Steel Production**
J.P. Hart — **Electroanalysis of Biologically Important Compounds**
T.F. Hartley — **Computerized Quality Control: Programs for the Analytical Laboratory**
Saad S.M. Hassan — **Organic Analysis using Atomic Absorption Spectrometry**
M.H. Ho — **Analytical Methods in Forensic Chemistry**
Z. Holzbecher, L. Diviš, M. Král, L. Šůcha & F. Vláčil — **Handbook of Organic Reagents in Inorganic Chemistry**
A. Hulanicki — **Reactions of Acids and Bases in Analytical Chemistry**
David Huskins — **General Handbook of On-line Process Analysers**
David Huskins — **Quality Measuring Instruments in On-line Process Analysis**
J. Inczédy — **Analytical Applications of Complex Equilibria**
Z.K. Jelínek — **Particle Size Analysis**
M. Kaljurand & E. Küllik — **Computerized Multiple Input Chromatography**
R. Kalvoda — **Operational Amplifiers in Chemical Instrumentation**
I. Kerese — **Methods of Protein Analysis**
S. Kotrlý & L. Šůcha — **Handbook of Chemical Equilibria in Analytical Chemistry**
J. Kragten — **Atlas of Metal-ligand Equilibria in Aqueous Solution**
A.M. Krstulović — **Quantitative Analysis of Catecholamines and Related Compounds**
F.J. Krug & E.A.G. Zagatto — **Flow Injection Analysis in Agriculture & Environmental Science**
V. Linek & V. Vacek — **Measurement of Oxygen by Membrane–Covered Probes**
C. Liteanu & S. Gocan — **Gradient Liquid Chromatography**
C. Liteanu & I. Rică — **Statistical Theory and Methodology of Trace Analysis**
Z. Marczenko — **Separation and Spectrophotometric Determination of Elements**
M. Meloun, J. Havel & E. Högfeldt — **Computational Methods in Potentiometry and Spectrophotometry**
O. Mikeš — **Laboratory Handbook of Chromatographic and Allied Methods**
J.C. Miller & J.N. Miller — **Statistics for Analytical Chemistry**
J.N. Miller — **Fluorescence Spectroscopy**
J.N. Miller — **Modern Analytical Chemistry**
J. Minczewski, J. Chwastowska & R. Dybczyński — **Separation and Preconcentration Methods in Inorganic Trace Analysis**
T.T. Orlovsky — **Chromatographic Adsorption Analysis**
D. Perez-Bendito & M. Silva — **Kinetic Methods in Analytical Chemistry**
B. Ravindranath — **Principles and Practice of Chromotography**
V. Sedivec & J. Flek — **Handbook of Analysis of Organic Solvents**
R.M. Smith — **Derivatization for High Pressure Liquid Chromatography**
R.V. Smith — **Handbook of Biopharmaceutic Analysis**
K.R. Spurny — **Physical and Chemical Characterization of Individual Airborne Particles**
K. Štulík & V. Pacáková — **Electroanalytical Measurements in Flowing Liquids**
O. Shpigun & Yu. A. Zolotov — **Ion Chromatography in Water Analysis**
J. Tölgyessy & E.H. Klehr — **Nuclear Environmental Chemical Analysis**
J. Tölgyessy & M. Kyrš — **Radioanalytical Chemistry**
J. Urbanski, *et al.* — **Handbook of Analysis of Synthetic Polymers and Plastics**
M. Valcárcel & M.D. Luque de Castro — **Flow-Injection Analysis: Principles and Applications**
C. Vandecasteele — **Activation Analysis with Charged Particles**
J. Veselý, D. Weiss & K. Štulík — **Analysis with Ion-Selective Electrodes**
F. Vydra, K. Štulík & E. Juláková — **Electrochemical Stripping Analysis**
N. G. West — **Practical Environment Analysis using X-Ray Fluorescence Spectrometry**
F.K. Zimmermann & R. E. Taylor-Mayer — **Mutagenicity Testing in Environmental Pollution Control**
J. Zupan — **Computer-Supported Spectroscopic Databases**
J. Zýka — **Instrumentation in Analytical Chemistry**

NUCLEAR ENVIRONMENTAL CHEMICAL ANALYSIS

JURAJ TÖLGYESSY
Professor of Nuclear and Environmental Chemistry
Slovak Technical University
Bratislava, Czechoslovakia

and

EDWIN H. KLEHR
Professor of Environmental Science
University of Oklahoma, Norman, USA

ELLIS HORWOOD LIMITED
Publishers · Chichester

Halsted Press: a division of
JOHN WILEY & SONS
New York · Chichester · Brisbane · Toronto

First published in 1987 by
ELLIS HORWOOD LIMITED
Market Cross House, Cooper Street,
Chichester, West Sussex, PO19 1EB, England
The publisher's colophon is reproduced from James Gillison's drawing of the ancient Market Cross, Chichester.

Distributors:
Australia and New Zealand:
JACARANDA WILEY LIMITED
GPO Box 859, Brisbane, Queensland 4001, Australia
Canada:
JOHN WILEY & SONS CANADA LIMITED
22 Worcester Road, Rexdale, Ontario, Canada
Europe and Africa:
JOHN WILEY & SONS LIMITED
Baffins Lane, Chichester, West Sussex, England
North and South America and the rest of the world:
Halsted Press: a division of
JOHN WILEY & SONS
605 Third Avenue, New York, NY 10158, USA

British Library Cataloguing in Publication Data
Tölgyessy, J.
Nuclear environmental chemical analysis. —
(Ellis Horwood series in analytical chemistry).
1. Chemistry, Analytic 2. Environmental chemistry
3. Radiochemical analysis
I. Title II. Klehr, Edwin H.
543'.088 QD75.3

Library of Congress Card No. 87–2726

ISBN 0–7458–0176–5 (Ellis Horwood Limited)
ISBN 0–470–20834–1 (Halsted Press)

Phototypeset in Times by Ellis Horwood Limited
Printed in Great Britain by R. J. Acford, Chichester

Table of contents

Preface

One of the main characteristics of our natural environment is the fact that chemical processes and quality play a crucial role in determining its behaviour. Moreover, because they are highly coupled with the biosphere and therefore with the well-being of humanity, it is mandatory that we have as complete an understanding of them as possible.

This understanding presents an unfinished and difficult task, however. It requires the finest chemical measurements, and therefore the continual development of precise and extremely sensitive analytical methods of all types — chemical, physico-chemical and, more recently, nuclear techniques.

The aim of the book is to describe the application of nuclear analytical methods to chemical analysis of the environment in the broadest sense, and hence the title "nuclear environmental chemical analysis". We wish to present the basic concepts and principles of various nuclear methods, including sampling, sample preservation and sample preparation, as well as their advantages and limitations for environmental monitoring.

We believe this to be the first book in the literature which describes *all* known nuclear analytical methods which are or may become useful in environmental studies, environmental science, engineering and technology, but we hope it will not be the last.

One of us (EHK) wishes to thank the Slovak Ministry of Education, the Slovak Technical University and the International Research and Exchanges Board for their generous support which greatly helped to make this book possible.

THE AUTHORS

1

The philosophy of nuclear environmental chemical analysis

1.1 THE IMPORTANCE OF OUR ENVIRONMENT

The chemical analysis of environmental systems is of crucial importance and a challenge to all who are interested in the development and application of nuclear analytical techniques. Such analysis is crucial because of our great need for a holistic understanding of the condition, behaviour and control of our environment. Never before have we been faced with the real possibility that ignorance and apathy might lead to drastic and long-term changes that could make life as we know it no longer possible on our planet. It is a challenge because of the complexity of environmental systems, and hence their analysis requires a great deal of sophistication in the understanding of nuclear techniques and of their proper application.

Attitudes towards our natural environment have changed radically in the last few years, and this can be seen with respect to water resources, for example. Not so long ago, the public view was simply that water was a free gift of nature; we could use it in whatever way we wished, and the environment would somehow take care of itself and our wastes. Pollution was merely something that had a detrimental effect on our immediate use. There was little concern about the adverse effects of pollutants on the environment, and waste was something we needed to get rid of as easily and cheaply as possible. The emphasis was entirely on convenience and cost, with little interest in the future or protection of the environment. In other words, the solution to pollution was dilution.

In the 1960s a breakthrough occurred in our thinking. We suddenly became aware of our planet as a closed system; it does not have an infinite capacity for our wastes, and there is not an endless supply of fresh water or other sources. Moreover, the ability of our biosphere to support the human race can become so damaged that progress in the environmental sense, if not life itself, is no longer possible. One group's source is literally another's waste. A pollutant is now seen as something that can have a detrimental effect on a natural biogeochemical cycle or process. We are now concerned about the operation of the enviroment, and about the behaviour, fate and impact of pollutants. We are slowly accepting the fact that prevention is better than cure. As some say, a pollutant has become "a resource out of place".

It is, however, quite a different matter to talk about a pollutant as something with a direct effect on human health, or as something that may cause a long-term effect in the environment. Compare three examples: arsenic in drinking water, nutrients released by a sewage treatment plant, and acid rain. The first seems quite straightforward; maximum permissible concentrations can be set relatively easily. That is not so easy in the second case because of lack of understanding of cause and effect relationships regarding eutrophication, and very difficult in the third. We are still arguing loudly about the latter two cases. What kind of measurements should we make in the case of acid rain and how should they be interpreted? What really *is* the long-term effect? How can the problem be managed most effectively? Clearly, our understanding of the environment must be holistic and sophisticated, and requires the finest of chemical analysis.

1.2 THE SCOPE OF NUCLEAR ENVIRONMENTAL CHEMICAL ANALYSIS

By environmental analysis, we mean the holistic study of an environmental system, its component parts, their inter- and intra-relationships, and its interactions with its surroundings. Chemical environmental analysis is therefore much more comprehensive than is usually implied by the term "analytical chemistry". It must include: identification of chemical species, the determination of concentrations or activities, and the study of reactions or processes. *Nuclear environmental chemical analysis* is that applied science which uses radioactive materials and/or nuclear reactions to assist in the task of chemical analysis. Therefore, its scope is quite broad, but it can be presented by a simple diagram as shown in Fig. 1.1.

The scope of nuclear environmental chemical analysis has three main branches.

Analytical measurements (nuclear environmental analytical chemistry), which includes the development and application of nuclear analytical techniques for the identification of chemical species and the determination of their concentrations or activities. Also included are the measurements of certain chemical/physical and chemical/biological parameters, such as the concentration of suspended particulates in air or water, measurement of the moisture content of soils, etc.

Study of reactions or processes, which involves the use of nuclear measurement techniques for the study of chemical reactions or processes in the environment. Included are identification of a process or reaction, its

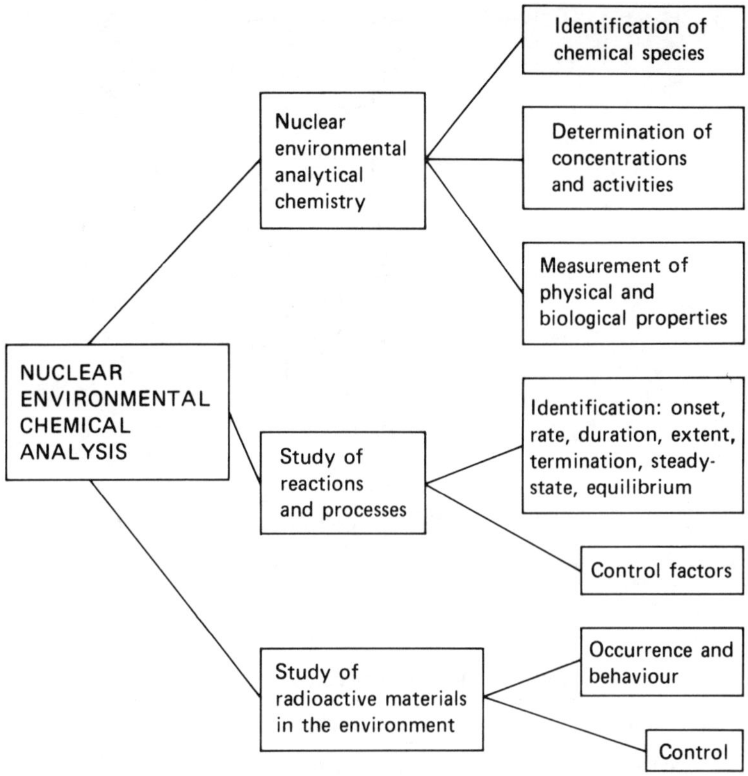

Fig. 1.1 — The scope of nuclear environmental analysis.

onset, rate, extent, duration, termination, steady-state and equilibrium conditions. Equally important is the elucidation of all the control factors for these various aspects, such as temperature, time, pH and redox potential.

Study of radioactive materials, which includes the use of nuclear analytical techniques to study the nature, occurrence, behaviour and control of natural and man-made radioactive materials in our environment.

In order that we may apply nuclear techniques in an optimum way, it is important to have a comprehensive awareness of the complexity of chemical reactions or processes in the environment. They are highly coupled and inter-related in various ways, as can be seen in Fig. 1.2. The natural environment can be thought of as including three main compartments: atmosphere (A), hydrosphere (H) and lithosphere (L). Each of these consists of a biotic (b) and an abiotic (a) subcompartment. The arrows indicate all possible reactions or processes within and between each subcompartment. One possible list of these is given in Table 1.1. It is of interest to note that, while the environment is indeed complex, the list of reactions or processes is not endless; the same ones appear again and again. The main causes of complexity are the great amount of coupling, and the subtlety of various control factors.

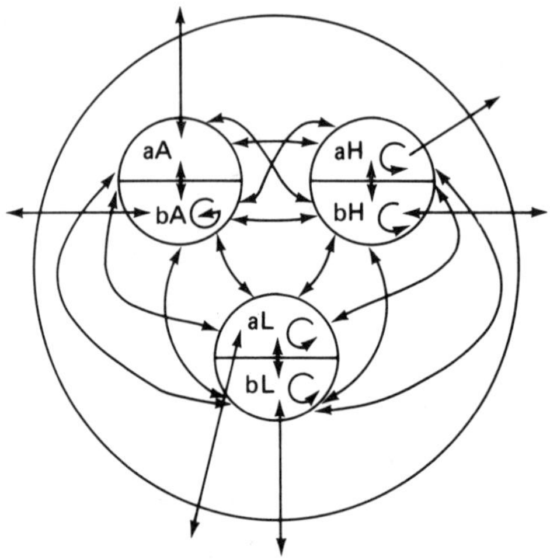

Fig. 1.2 — Reactions and processes in the environment.

Table 1.1 — Reactions and processes in the natural environment.

Chemical	Chemical/physical	Chemical/biological
Acid/base	Inflow/outflow	Uptake
Redox	Diffusion/mixing	Feeding
Complex formation	Settling/suspension	Metabolism
Ion-pair formation	Evaporation/precipitation	Elimination
Precipitation/dissolution	Coagulation/dispersion	
Ion-exchange	Gas transfer	
Sorption/desorption	Spray	
Photolysis	Dilution	
Rearrangement	Fall-out/wash-out	
	Storage	

1.3 THE NEED FOR VARIOUS KINDS OF CHEMICAL ANALYSIS

The role of and need for various kinds of chemical analysis can be indicated by a simple decision diagram such as that in Fig. 1.3 [1]. It is intended primarily for use with water resources, but is applicable to other areas. Water is a finite resource, and its availability can be the determining factor in the behaviour and quality of the environment, as well as in the economic and social well-being of a human community. It is necessary for various human needs, and analytical measurements are required to relate these needs to a suitable water quality, which in turn leads to water quality standards. An example of standards for drinking water in the USA is given in Table 1.2 [2].

To satisfy these needs, we use various sources of water, and analytical measurements must be used to evaluate their quality. If the standards and quality do not match, treatment is necessary, and a set of analytical

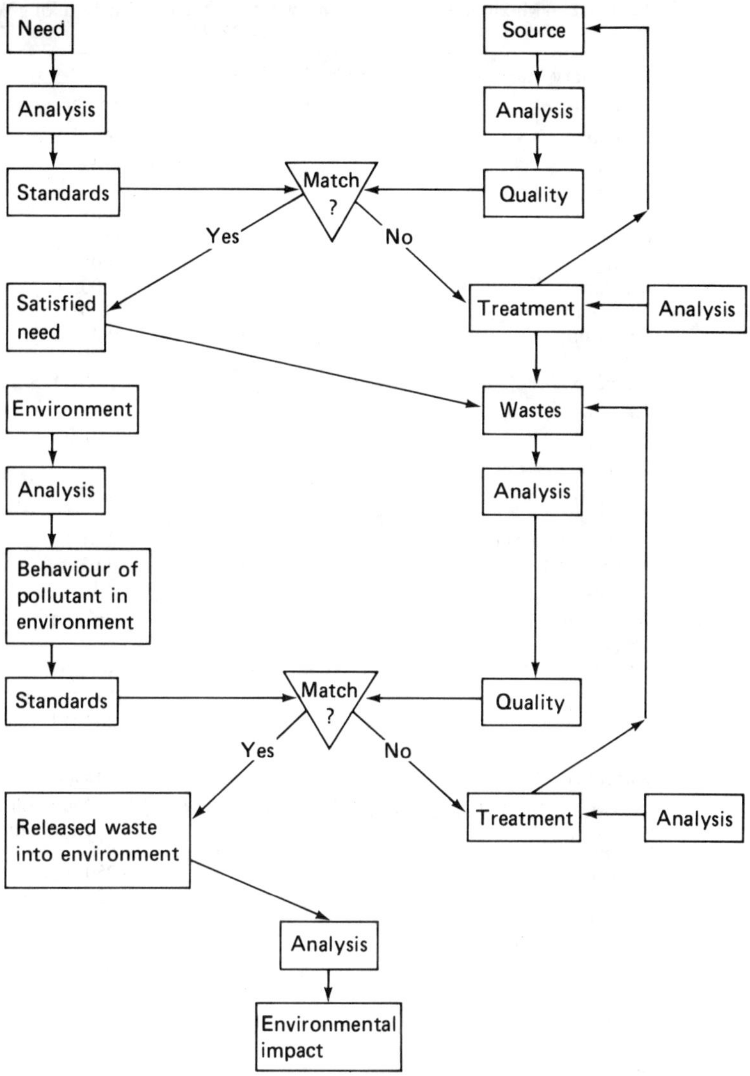

Fig. 1.3 — Decision diagram [1].

measurements is used to design, optimize and monitor the treatment system [3, 4].

Treatment and use generate wastes, and these must be disposed of in such a manner as to protect future reuse of the water, and also the environment. Hence, analytical measurements are needed to help understand the functioning of the environment and the behaviour and impact of various pollutants [5–8]. This understanding leads to a second set of standards [9]. Additional analytical measurements are needed to evaluate the characteristics of the wastes so that a decision can be made regarding

Table 1.2 — Maximum concentration levels from the US Safe Drinking Water Act (Public Law 93–523) [2].

1. Primary Maximum Concentration Levels (PMCL)	
Contaminant	PMCL (mg/l.)
Inorganic	
Arsenic	0.05
Barium	1.0
Cadmium	0.01
Chromium	0.05
Lead	0.05
Mercury	0.002
Nitrate (as N)	10.0
Selenium	0.01
Silver	0.05
Fluoride	1.4–2.4
Organic	
Endrin	0.0002
Lindane	0.004
Toxaphene	0.005
2,4-D	0.1
2,4,5-TP (Silvex)	0.01
Methoxychlor	0.01
TTHM	0.1
Radiological	(pCi/l.)
^{226}Ra	5
^{228}Ra	5
Gross α-activity	15 (excluding Rn and U)
^{3}H	20000
^{90}Sr	8
Turbidity	1 turbidity unit
2. Secondary Maximum Concentration Levels (SMCL)	
Contaminant	SMCL
Chloride	250 mg/l.
Colour	15 colour units
Corrosivity	Non-corrosive
Foaming agents	0.5 mg/l.
Iron	0.3 mg/l.
Manganese	0.05 mg/l.
Odour	3 TON
pH	6.5–8.5
Sulphate	250 mg/l.
Total dissolved solids	500 mg/l.
Zinc	5 mg/l.

disposal. If the waste is not compatible with the receiving environment, treatment is necessary, and more measurements are required for design, optimization and monitoring [10, 11].

When the waste has been suitably treated, it is released, but this is not the end of our concern. We must use a host of analytical measurements to monitor the response of the environment to the treated waste in order to help prevent any unforeseen detrimental effects, and to make the treatment as cost-effective as possible.

We see that analytical measurements of many kinds are needed for quite diverse purposes:

(1) establishment of standards related to intended use;
(2) evaluation of the quality of resources;
(3) design, optimization and monitoring of treatment of processing systems;
(4) study of the nature and behaviour of our environment;
(5) characterization of wastes;
(6) determination of the behaviour, fate and impact of pollutants in our environment;
(7) evaluation of the impact of treated wastes (and other human activities) on the environment over longer periods of time.

1.4 SURVEY OF TECHNIQUES AVAILABLE FOR CHEMICAL ANALYSIS OF THE ENVIRONMENT

Historically, a large number of analytical techniques have been used for chemical analysis of the environment. A complete discussion of all of them is beyond the scope of this book, but it is useful to give a brief survey of the different possibilities and approaches. It is impossible to give a definitive list because of confusion and overlap in terminology, rapid developments and disagreements as to syntax. Broadly speaking, we can identify categories of techniques as shown in Fig. 1.4 [12–15].

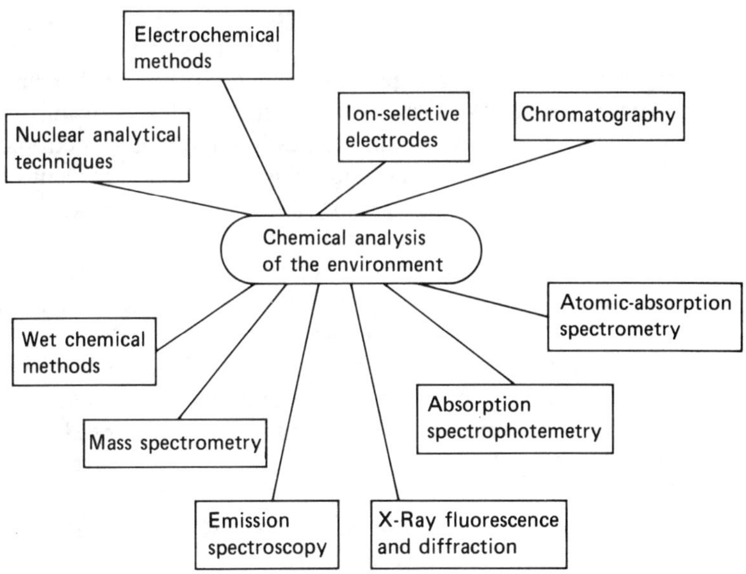

Fig. 1.4 — Analytical techniques used in chemical analysis of the environment.

Wet chemical methods generally involve rather simple procedures and equipment. Major disadvantages are inadequate sensitivity for most analytes of interest and sometimes lengthy analysis times. Examples include titrations and gravimetry.

Electrochemical methods. These techniques are based on either the measurement of a potential (potentiometry) or of a current (voltammetry). A good example of potentiometry is the use of ion-selective electrodes to measure species such as fluoride, chloride, bromide, iodide, nitrate, cyanide, thiocyanate, sulphide, cadmium, copper, lead, silver, calcium and potassium. Advantages include convenience and speed. Disadvantages are calibration problems, interferences, salt effects and limited applicability. Moreover, the measured signal is proportional to the *logarithm* of an activity, so small errors in voltage measurement may lead to relatively large errors in the calculated values. In samples of high ionic strength (e.g. brines, ocean water), the relationship between activity and concentration must be taken into account, which may be difficult. Further, the sensor responds only to the free ion, and not to complexes.

Voltammetric techniques include d.c., a.c. and differential-pulse polarography, fast linear-sweep voltammetry and anodic-stripping voltammetry. Classical (d.c.) polarography seldom has the sensitivity required for environmental measurements, but more recent variations such as the differential-pulse and linear-sweep techniques have improved the sensitivity in some cases. Anodic-stripping voltammetry has been used for some metals (Cu, Pb, Zn, Cd) in the sub-μg/ml range in water. Membrane-covered voltammetric electrodes (e.g. Ag cathode, Pb anode) are commonly used for the measurement of oxygen.

Chromatography represents a group of separation techniques, and must be combined with suitable detectors or measurement devices. An outstanding example is the use of gas chromatography for the determination of a wide variety of organic species. Detectors include devices using thermal conductivity, electron capture, flame ionization and mass spectrometry. The main advantages of gas chromatography in environmental analysis are the ability to separate a great variety of organic species from mixtures of various sorts and the availability of sensitive detectors. Among the disadvantages are expensive equipment and very demanding techniques. In practice, special laboratory facilities and highly skilled operators are required.

High-performance liquid chromatography (HPLC) uses a liquid mobile phase and a column of very small solid particles as the stationary phase. High pressures are necessary to get useful flow-rates, but high resolution is possible in favourable cases. The main advantage of HPLC is that samples can be analysed in the liquid phase and hence conversion into a gaseous phase is not necessary. Unfortunately, the detectors currently available (refractive index, UV absorption, fluorescence) are generally not as sensitive as those used for gas chromatography.

Mass spectrometry is sometimes used in conjunction with gas chromatography for the measurement of organic species. It can provide both qualitative (identification) and quantitative information. Computerized data banks are available, containing information about a large number of compounds.

In favourable cases, spark-source mass spectrometry has been used for elemental analysis at trace levels. Though mass spectrometry can be a valuable addition to the list of analytical techniques, its applicability is seriously limited by the need for expensive equipment, special laboratory facilities and highly trained personnel.

Absorption spectrophotometry refers to techniques which utilize the absorption of light by suitable chemical species. In essence, it consists of measuring the amount of light (preferably monochromatic) passing through a solution containing the analyte or some derivative. Examples include the determination of ammonia, nitrate and nitrite in water samples, and sulphur dioxide in air. Advantages include broad applicability and less expensive equipment. Some disadvantages are interference problems, inadequate sensitivity for some metals and sometimes cumbersome, rather irreproducible procedures.

Atomic-absorption spectrometry (AAS) is based on the absorption of virtually monochromatic light by a cloud of atoms of the analyte. The light is produced by emission from the same kind of atoms as those being determined. This means that exactly the same wavelength of light is produced as that being absorbed, and hence a high degree of selectivity is possible. Hollow-cathode lamps containing the elements of interest are used as light sources. Atomizers include high-temperature flames (e.g. O_2–C_2H_2) and graphite furnaces. As a special case, mercury is measured at ordinary temperature by reduction to the elemental state, followed by sweeping the mercury atoms into a measuring cell with air.

AAS has become the preferred method for most metals in environmental analysis. Some advantages are good sensitivity, speed and the possibility of multielement analysis. Disadvantages are relatively expensive equipment and the need for well trained personnel.

Emission spectroscopy is comprised of methods based on the observation of spectral lines emitted during deactivation of analyte atoms or ions from an excited state to a state of lower energy. Excitation sources for conventional units include flames (flame photometry), and high-voltage discharge from graphite electrodes. It has been used for many metals, particularly with atmospheric and biological samples.

A breakthrough in emission spectroscopy was the development of inductively-coupled plasma emission spectroscopy (ICP). The excitation source is an incandescent plasma of argon, heated by radiofrequency induction. Extremely high temperatures can be produced (10^4 K) which makes possible the simultaneous measurement of as many as 30 elements, including important semi-metals and non-metals such as As, B and Se. Interferences are less bothersome, and the sensitivity is generally better than with flame excitation.

X-Ray fluorescence is based on the detection of X-rays emitted when electrons fall into inner vacancies in analyte atoms. These vacancies are created by bombardment of the sample with high-energy sources (e.g. X-rays, γ-rays, protons). The spectra produced yield qualitative and quantitative information; emission wavelengths may be used to identify elements, and intensities to measure amounts. The X-ray spectra produced can be measured either by wavelength-dispersive or energy-dispersive (e.g. Si/Li detector) systems.

Currently, X-ray fluorescence seems particularly well suited for the study of air pollutants. The use of radioactive materials as excitation sources is a recent development, and may lead to the availability of useful field equipment.

Advantages include the possibility of multielement analysis and field work. Among the disadvantages are matrix problems and lack of sensitivity in environmentally important cases.

1.5 NUCLEAR ANALYTICAL TECHNIQUES

Nuclear analytical chemistry is a part of applied radiochemistry and analytical chemistry. It is comprised of those analytical methods which use the nuclear characteristics of appropriate nuclides to gain qualitative and quantitative information about various substances and about our environment. The common feature of all such methods is the detection and measurement of nuclear radiation and/or characteristic X-rays. Two basic situations are included: the use of internal and external radiation. *Internal radiation* is emitted by the sample itself or a radioactive isotope of an element in the sample which may be originally present, added or produced by activation. *External radiation* is produced external to the sample, and interacts with it by absorption, scattering, capture and thermal effects. Generally, changes in external radiation caused by sample interactions are measured.

Both qualitative and quantitative measurements can be made, the latter being more commonly used. The basis of all the quantitative measurements is the direct proportionality between the amount of radioactive substance in the sample (or the amount of radioactive isotope used to indicate a non-radioactive analyte) and the intensity of the measured radiation. In most cases, these are relative or comparative measurements. Fortunately, absolute measurements of an amount of radioactivity are seldom necessary.

Less frequently, qualitative measurements are made in which nuclides are identified on the basis of their half-life and the nature and energy of the emitted radiation.

The various nuclear analytical methods available can be classified as in Fig. 1.5 [16]. There are two major groups: direct and indirect methods.

Direct methods involve the direct measurement of a radionuclide in an environmental sample (e.g. ^{131}I in milk, ^{90}Sr in bones, ^{137}Cs in soils, ^{238}U in ores).

The largest group of *indirect methods* is comprised of *indicator methods* in which a radioactive nuclide is added to the sample (usually in known amounts), and is used to indicate the presence of an analyte. Examples of indicator methods include isotope dilution analysis, radiometric titrations and radio-reagent analysis. These methods are usually simpler than *activation methods*, in which radionuclides are produced in the sample by some activation technique.

Non-activation interaction analysis is based on the absorption and/or

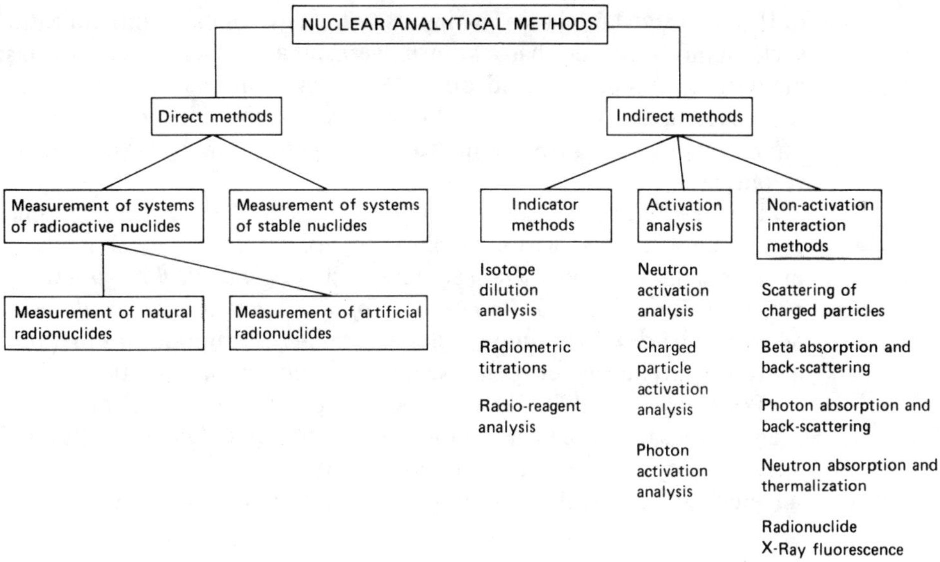

Fig. 1.5 — Nuclear analytical methods [16].

scattering of alpha or beta particles, photons and neutrons, during their interaction with the sample.

Nuclear environmental chemical analysis, and specifically the development and application of nuclear analytical methods, represents an interesting positive-feedback loop, as pictured in Fig. 1.6, modified from the work

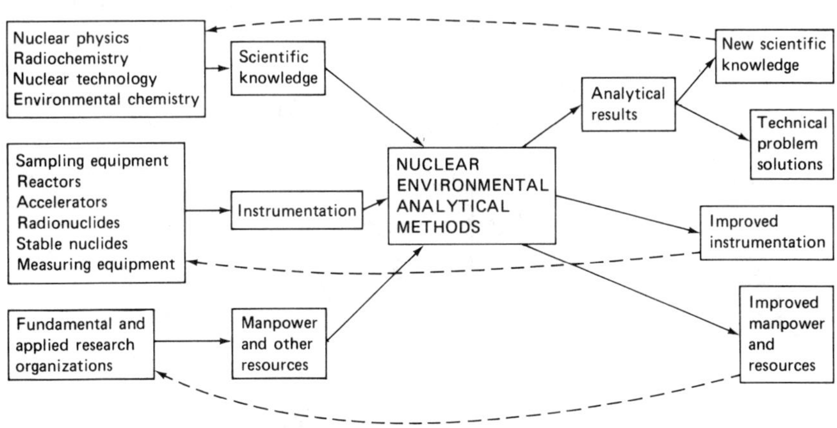

Fig. 1.6 — The positive-feedback mechanism for the development of nuclear environmental analytical methods (adapted from [17]).

of Bujdosó and Tóth [17]. The inputs to the development and utilization of such methods include: basic knowledge from a number of sciences, instrumentation, manpower, and other resources. Outputs include: analytical results leading to new scientific knowledge and solutions to problems, and improvements in instrumentation, personnel capabilities, and other resources.

Braun [18] has presented an interesting study of the general trends in recent publications on the applicability of trace analytical methods to environmental samples. Data sources included *Analytical Abstracts* for the period 1975–1984, and *Proceedings of the International Conference on Modern Trends in Activation Analysis*, and his assumption was that there is a strong correlation between research and publication activity.

For the period 1975–1984, publication production in the various instrumental areas was ranked as follows: chromatography (gas and HPLC) 33%, spectrometry 31%, electroanalysis (primarily ion-selective electrodes) 20% and nuclear analytical techniques 16%. This ranking corresponded well with

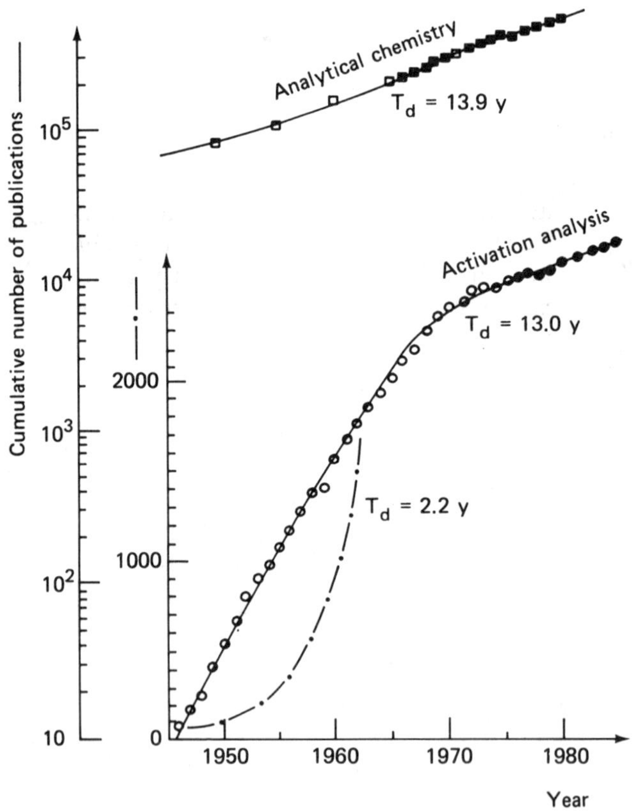

Fig. 1.7 — Cumulative growth rate of publications on activation analysis and on analytical chemistry, 1936–1985. (Reproduced from [18] by courtesy of Professor T. Braun).

that Braun found by taking an opinion poll of editorial boards of the major analytical chemistry journals.

Activation analysis was by far the most frequently used technique for the analysis of inorganic trace pollutants. He ranked the various activation methods as follows: thermal neutrons, fast neutrons, charged particles and photons. Of significance is the fact that the growth rate of publications on activation analysis has decreased substantially, possibly owing to the small number of appropriate reactor facilities. During the period 1936–1970, its doubling time was only 2.2 years, but this has since slowed to 13.0 years, which is close to the 13.9 years for general analytical chemistry (see Fig. 1.7 [18]).

He also pointed out that in many cases a preconcentration step was necessary in order to achieve the sensitivity required for environmental analysis. He ranked the more commonly used preconcentration techniques in the following descending order: liquid–liquid extraction, precipitation/co-precipitation, ion-exchange, and chelate exchange.

Braun speculated on the potential usefulness of instrumentation, based on nuclear techniques, that can be used in the field for direct measurements on the environment. He coined the term "nucleotode" for such devices, and indicated that preliminary results suggested such sensors can be built, for example in combination with X-ray fluorescence analysis.

1.6 SAMPLING AND SAMPLE PRESERVATION

A laboratory filled with the finest equipment, the most skilled personnel and using the best procedures will not produce reliable information about our environment unless adequate attention is given to the matter of sampling. Results cannot be better than the samples from which they are obtained, and sampling is often the weakest link in the analysis chain. This obvious truth is always agreed upon, but many times overlooked or even ignored in practice.

On the other hand, it must be admitted that sampling and the handling of samples is not an easy matter; it can be subtle, complex and frustrating, and this is particularly true when dealing with the environment. It is impossible to list a set of rules, procedures, equipment, etc. that will always and everywhere guarantee reliable results, and to a large degree the matter has not been given sufficient study.

The general principle is quite clear and easy to state. A sample must adequately represent, in a spatial and temporal sense, the environmental system from which it was taken. Its nature must not interfere with the subsequent measurements or interpretation of the results.

In any sampling programme, the following items must be considered.

(1) Reasons for taking the samples.
(2) Type of sample to be taken.
(3) Location of sampling points.
(4) Size and number of samples.
(5) Frequency and duration of sampling.

(6) Sampling equipment and techniques.
(7) Samnple containers.
(8) Sample preservation.
(9) Transport and holding times.
(10) Documentation.

In general, the number and location of sampling points depends on the spatial homogeneity of the system under study. Unfortunately, the homogeneity may not be known ahead of time, and preliminary sampling may be necessary. Similarly, the frequency and duration depend on the temporal homogeneity of the system. The more rapidly a system changes, the more frequently samples need to be taken, and ideally, a system would be sampled continuously, but this is often impractical.

The size of sample generally depends on the analytical techniques, the transportation system, and the need for replicate measurements. Equipment, containers, preservation techniques and holding times depend on the type of sample taken and the nature of the analytes.

Adequate documentation is of obvious importance, but is easily overlooked. It should include all information necessary for unambiguous correlation of final analytical results with the original sample, and also all information that could have a bearing on the interpretation of data obtained from the sample.

The general principles of environmental sampling apply to samples taken for the measurement of radioactive constituents, and a more detailed discussion of this very important topic is given in Chapter 2.

1.7 ANALYSIS OF DATA

Regardless of whether a nuclear analytical method is based on relative or absolute measurements, results must be evaluated according to agreed criteria. The literature indicates that, while there is a concensus in principle, a considerable confusion exists as to the proper definitions and evaluation of these criteria.

For example, there seems to be some ambiguity regarding the essential concepts of accuracy, precision, reproducibility and sensitivity [19, 20].

By *accuracy* we mean the closeness of agreement between a true value and the mean result obtained by applying the same experimental procedure to the same sample a large number of times. The smaller the systematic part of the errors (bias) which affect the results, the more accurate the procedure. Since "true" values are never known with certitude, statements of accuracy are never absolute.

Precision is a statement of the closeness of agreement among the results obtained by appling the same experimental procedure to the same sample several times. The smaller the random part of the experimental errors, the more precise is the procedure. Some function of the standard deviation is commonly used to state precision values.

It should be obvious that accuracy and precision are not the same. A procedure with a large bias (systematic error) can be quite precise, but will be inherently inaccurate. On the other hand, if it has genuine accuracy, it will generally have good precision as well. *Reproducibility* and precision seem to have the same meaning.

In any analytical method, the accuracy and precision of the final data depend on all steps of the procedure, and not only on the final measurement. Hence, any statements regarding these criteria should also include a discussion of the contribution to them from each step in the process. An attempt should be made to evaluate all sources of random errors and of bias (systematic errors).

The term *sensitivity* seems particularly troublesome, and has been given various meanings by different authors. It can refer to the minimum amount of analyte that will produce a signal which can be distinguished from the noise in the final measurement with a certain degree of confidence. In this sense, it depends only on the final measurement technique. If, however, it refers to the minimum amount of concentration of the analyte that can be detected in the sample, then it will depend not only on the final measurement technique (signal *vs.* noise) but will also be affected by the entire analytical procedure, including choice of sample size, preconcentrations, dilutions, yields and recoveries. Hence, it is clear that any quoted value of sensitivity should include a statement as to how it was defined and calculated. Further confusion arises from the definition of sensitivity as the slope of a calibration graph (or linear response function [21]. For example, in instrumental neutron activation analysis (INAA), sensitivity is commonly defined as the number of counts recorded per unit weight of an element under prescribed conditions.

Other terms are sometimes used, such as *lower limit of detection*, the smallest amount or concentration of a radionuclide that can be distinguished in a sample by a given measurement system in a preselected counting time at a given confidence level.

It is of interest to note that nuclear analytical procedures are one of those rare cases where precision and sensitivity can be somewhat controlled by manipulation of measuring times (i.e. counting times). All other things being equal, longer counting times can reduce the relative standard deviation of a measurement result (counts) and hence improve the overall precision of the method.

In general, the analysis of data taken from nuclear analytical techniques is the same as for other methods [12]. However, there are some significant differences. The analysis of data in these methods frequently includes the statistics of radioactive decay. In most cases, straightforward and well documented techniques can be used, such as the Gaussian (normal) or Poisson distributions, and conventional error propagation theory [22].

1.8 COMPARISON OF ANALYTICAL METHODS

In specialty analytical laboratories, which are operated for specific analysis problems, the nature of the samples and analytes are generally known in advance; moreover, variations in sample matrix and concentrations are usually within a relatively small range. Therefore, the selection of methods and the design laboratory procedure are comparatively straightforward, and a limited number of procedures will usually suffice.

In environmental analysis however, it is often difficult to know beforehand what kind of samples will need to be analysed, the parameters which need to be measured and their expected ranges. In such a situation, it is necessary to be familiar with the applicability of various methods with respect to limits of detection, accuracy, precision, selectivity, reproducibility, interferences, speed and cost.

A large number of analytical methods are available for measurements on environmental samples, but no one of them provides a general solution; all have their advantages, limitations and usefulness. The sensitivity of nuclear analytical methods is compared to other methods in Table 1.3 [23], and it can be seen that these methods are as good as or better than others. In favourable cases nuclear methods have far superior sensitivity (values are quoted to the nearest order of magnitude). In Table 1.4 [24] some general characteristics of a variety of instrumental methods are presented. In Figure 1.8 [25, 26] the applicability of some important approaches is shown. Figure 1.9 [25, 26] shows the precision expressed as relative standard deviation, and Fig. 1.10 [25, 26] gives a comparison of the analysis time required for these methods. This information can be used for the comparison of analytical methods and for help in choosing a solution to a specific problem.

Table 1.3 — Detection limits of various analytical methods [23]. (Reproduced by permission of the copyright holders, SNTL, Prague).

Method		Detection limit (g)	Volume of sample (ml)	Concentration for an analyte of m.w. 100 (mole/l.)
Non-radioactive analytes				
Chemical	Classical methods	10^{-6}	10	10^{-6}
	Special ultramicro procedures	10^{-12}	0.001	10^{-8}
Physical-chemical	Absorption spectro-photometry	10^{-8}	1	10^{-7}
		10^{-9}	0.01	10^{-7}
	Inversion polarography	10^{-10}	10	10^{-10}
	Mass spectrometry	10^{-12}	0.01	10^{-9}
	Gas chromatography	10^{-13}	0.001	10^{-9}
	Atomic-absorption spectrometry	10^{-13}	0.01	10^{-10}
Nuclear	Substoichiometric IDA	10^{-11}	1	10^{-10}
	Neutron activation analysis	10^{-13}	1	10^{-12}
	Radioactive substances			
	Ionization and scintillation	10^{-19}	1	10^{-18}
	Micro radiography	10^{-22}	1	10^{-21}

Table 1.4 — Comparison of some instrumental analytical methods [24] (Reproduced by permission of the copyright holders, SNTL, Prague).

Method	Analyte	Conc. range*	Selectivity	Analysis time†	Costs§
Classical polarography	Ident. and detn. of about 80 cations of heavy metals; many organic substances	$1–10^{-4}$	Good	Short	Low–medium
Inverse polarography	Detn. of about 40 heavy metals heavy metals	$10^{-4}–10^{-10}$	Poor	Short	Medium
Atomic-absorption spectro-photometry	Detn. of many heavy metals	$10^{-2}–10^{-6}$	Good	Short	Medium
Molecular absorption spectrometry (UV, visible)	Detn. of practically all cations except alkali metals; a number of anions; some organic species	$1–10^{-5}$	Fair	Short	Medium
Infrared spectrometry	Ident. of functional groups and detn. of structure of org. substances	$1–10^{-1}$	Fair	Short	Medium
Emission spectrometry	Ident. and detn. of almost all elements (multi-element method)	$10^{-1}–10^{-5}$	Good	Short	Medium
Electron-microprobe	Ident. and detn. of elements with at. no. >5; local surface analysis	$1–10^{-3}$	Good	Short	High
NMR spectrometry	Ident.; structural anal.; detn. of mainly organic species	$10–10^{-1}$	Poor	Short	High
Electron spin resonance spectrometry	Ident. and structural anal. of species which have an odd no. of electrons	$10–10^{-1}$	Poor	Short	High
Mass spectrometry	Ident. of org. species; structural analysis; rel. masses; ident. and detn. of all isotopes	$10–10^{-1}$	Good	Medium	High
Neutron activation analysis	Multi-elemental ident. and detn.	$10^{-1}–10^{-7}$	Fair	Medium–long	High
Chromatography	Ident. and detn. of many organic and inorganic species	$10^{-1}–10^{-7}$	Good	Short	Medium
X-Ray fluorescence with X-ray tubes	Multielement detn. of elements with at. no. >15	$10^{-1}–10^{-52}$	Good	Short	High
X-Ray fluorescence with radionuclide excitation	Multielement detn. of elements with at. no. >11	$10–10^{-4}$	Good	Short	Low-medium
Classical IDA	Detn. of organic and inorganic species which can be labelled with appropriate radionuclides	$10^{-1}–10^{-6}$	Fair-good	Medium	Low
IDA-substoichiometric and sub-equivalence methods	Detn. of most elements when appropriate labelling nuclides are available	$10–10^{-8}$	Good	Medium	Low

* Percentage of analyte, sample weight 1 g.
† Short, sec–min; medium, hr; long, days.
§ Equipment cost: low <$10000; medium $10000–75000; high >$75000.

In general, the selection of an analytical method for a given problem should be based on at least three kinds of criteria: analytical, occupational and financial [27]. Included in the analytical criteria are the sensitivity, blank values, matrix effects, chemical yields, accuracy, precision, limits of detection, turnover-over time and sample-handling capacity. Occupational criteria involve the health and safety of exposed personnel. Financial criteria are generally based on cost per datum and long-term capital cost.

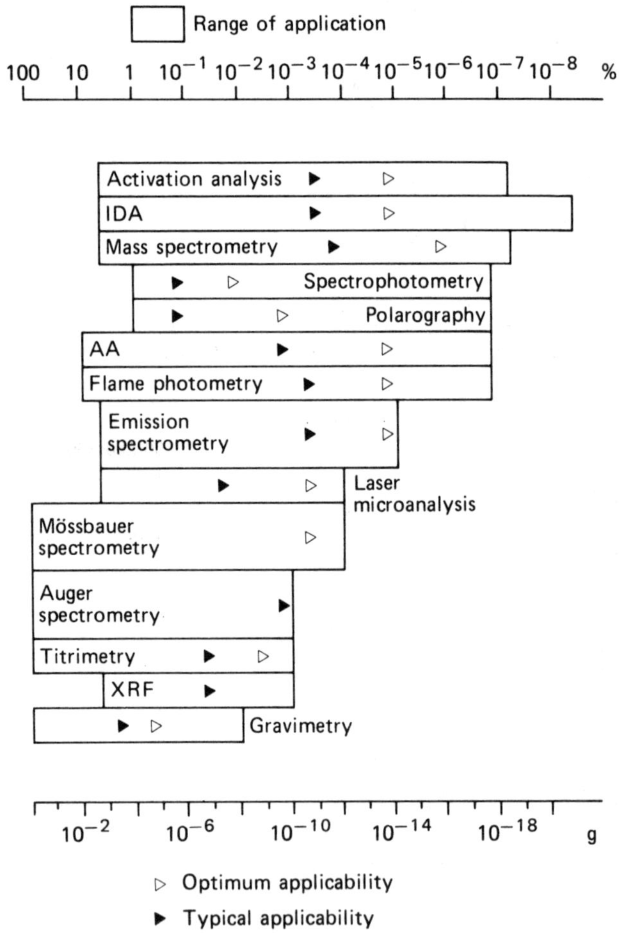

Fig. 1.8 — Range of application of some analytical methods. (Reproduced from [25, 26] by permission of the copyright holders.)

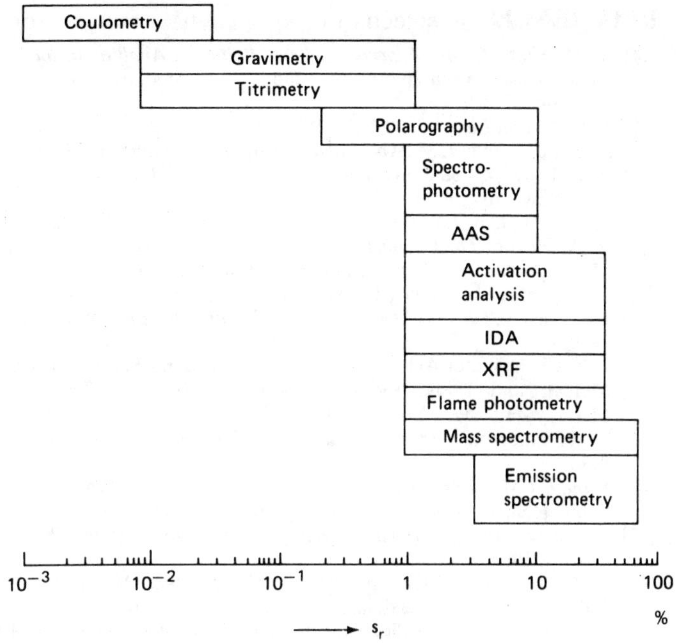

Fig. 1.9 — Precision of analytical methods, expressed as the relative standard deviation s_r. (Reproduced from [25, 26] by permission of the copyright holders.)

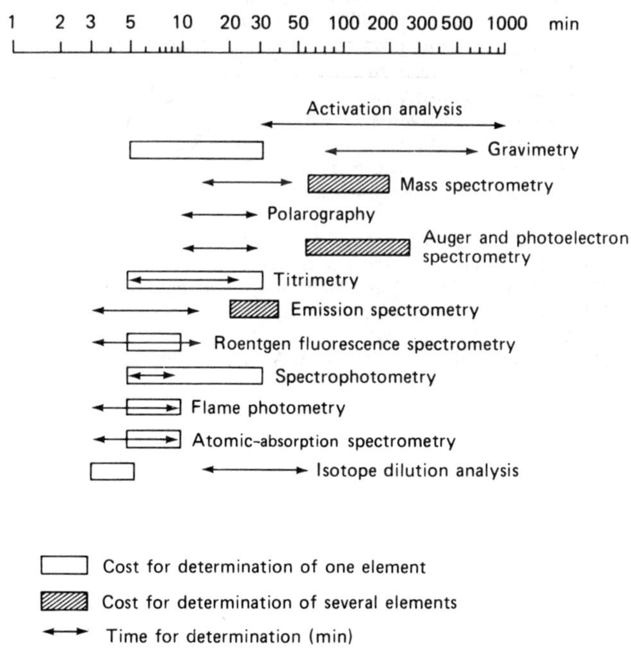

Fig. 1.10 — Analysis time and relative cost for different analytical methods. (Reproduced from [25, 26] by permission of the copyright holders.)

REFERENCES

[1] E. H. Klehr, *Some Aspects of Water Pollution Analysis in the United States*, Analyža látok znečistujucich životné prostredie, Nový Smokovec, Czechoslovakia, 22–23 April 1986, pp. 31–49.
[2] *U.S. Federal Register*, 42(132): 35764, 1977.
[3] S. D. Faust and O. M. Aly, *Chemistry of Water Treatment*, Butterworths, Boston, 1983.
[4] H. S. Peavy, O. R. Rowe and G. Tchobanoglous, *Environmental Engineering*, McGraw-Hill, New York, 1985.
[5] S. E. Manahan, *Environmental Chemistry*, 4th Ed., Willard Grant Press, Boston, 1984.
[6] R. A. Horne, *The Chemistry of our Environment*, Wiley, New York, 1978.
[7] M. Lippmann and R. B. Schlessinger, *Chemical Contamination in the Human Environment*, Oxford University Press, New York, 1979.
[8] I. J. Tinsley, *Chemical Concepts in Pollutant Behaviour*, Wiley–Interscience, New York, 1979.
[9] U.S. Clean Water Act, Public Law 92-500, *Federal Register*, 45 (231): 79318, 1980.
[10] Metcalf and Eddy, Inc., *Wastewater Engineering: Treatment, Disposal, Refuse*, McGraw-Hill, New York, 1979.
[11] J. W. Paterson, *Industrial Wastewater Treatment Technology*, Butterworths, Boston, 1985.
[12] *Standard Methods for the Examination of Water and Wastewater*, 16th Ed., American Public Health Association, Washington, D.C., 1985.
[13] J. Albaiges (ed.), *Analytical Techniques in Environmental Chemistry*, Pergamon Press, Oxford, 1982.
[14] A. C. Stern, *Air Pollution, Vol. III, Measurement, Monitoring and Surveillance of Air Pollution*, 3rd Ed., Academic Press, New York, 1976.
[15] A. L. Page, R. H. Miller and D. R. Keeney, *Methods of Soil Analysis*, ASA, SSSA, Madison, Wisc., 1983.
[16] J. Tölgyessy, *Nukleárna Ekoanalytická Chémia*, pp. 11–30. Analýza látok znečistujúcich životné prostredie, Nový Smokovec, Czechoslovakia, 22–23 April 1986.
[17] E. Bujdosó and L. Tóth, *J. Radioanal. Chem.*, 1980, **59**, 255.
[18] T. Braun, *Charting the Trends in Nuclear Techniques for Analysis of Inorganic Environmental Pollutants*, IAEA Consultants Meeting on Nuclear Techniques for Analysis of Environment Samples, Karlsruhe, FGR, 19–21 February 1986.
[19] International Organization for Standardization, *Statistics — Vocabulary and Symbols*, International Standards, ISO 3534, 1977.
[20] G. Hemon and J. C. Philippot, *A New Method for Processing INAA Results Without the Necessity of Standards*, IAEA Consultants Meeting on Nuclear Techniques for Analysis of Environmental Samples, Karlsruhe, FRG, 19–21 February 1986.
[21] IUPAC, *Pure Appl. Chem.*, 1976, **45**, 99.
[22] N. Tsoulfanidis, *Measurement and Detection of Radiation*, McGraw-Hill, New York, 1983.
[23] V. Majer, *Základy jaderné chemie*, SNTL, Prague, 1981.
[24] Z. Holzbecher, *Základy analytické chemie*, SNTL, Prague, 1979.
[25] K. Danzer, E. Than and D. Molch, *Analytik*, Akademische Verlag., Leipzig, 1976.
[26] J. Klas, J. Tölgyessy and J. Lesný, *Sub-superkvivalentová izotopová zriedovacia analýza*, Veda, Bratislava, 1985.
[27] H. A. Das, A. Faanhof, and H. A. van der Sloot, *Environmental Radioanalysis*, Elsevier, Amsterdam, 1983.

2

Environmental sampling

In general, sampling is the most important step in an entire analytical procedure, but this simple fact is easily overlooked (see Section 1.6). The objective of a sampling programme is to collect a representative portion of the environment, with due regard to the purpose of the study, and transfer it to a laboratory or field station while it still represents the current state of the system. Samples must represent the system in both a time and spatial sense, must be taken with the parameters to be measured borne in mind, and must be handled so that no detrimental changes occur before measurements are made. This necessitates having a good idea of the variation of the environment in terms of position and time, and taking every possible precaution to make the sampling as reproducible as possible.

It is not possible to give detailed procedures for all sampling programmes, because of the great diversity of circumstances, and ultimately the success of a sampling programme depends on the judgement of the analyst.

2.1 ATMOSPHERIC SAMPLING

Environmental air sampling is one of the most important and difficult steps in the surveillance of air pollution.

Particulate air-sampling techniques include filtration, electrostatic precipitation, thermal precipitation, gravitational settling, centrifugal separation, and impingement.

2.1.1 Filtration

The most popular air sampling method is *filtration*. Filters have high efficiencies, even for very small particles, and low capital costs. The filtration mechanism is direct interception and inertial impaction for the larger particles, and Brownian or diffusional impaction for the smaller ones. The sample is collected with a high-volume sampler (using a glass fibre or polystyrene filter), or a low-volume sampler (using a membrane or other suitable filter), or with certain other kinds of samplers.

The *high-volume air sampler* consists of an air-suction assembly, filter holder, flow-rate measuring assembly and shelter (Fig. 2.1). The particulate matter is generally collected by suction of air at a flow-rate of approximately 1.2–1.7 m^3/min [1].

The *low-volume air sampler* consists of a pump, filter holder and flow-rate measuring assembly (Fig. 2.2). The suction rate is usually 20–30 l./min for metal content analysis [1].

The suction requires a *source of reduced pressure*, such as a rotary vacuum pump, piston air pump, diaphragm pump, or suction fan, of which the rotary vacuum pump and fan are widely used for the sampling of particulate matter.

Air metering devices are needed to relate the quantity of particulate matter collected to the amount of air sampled. The devices widely used for this purpose are rate meters through which the air passes in a continuous stream, and volume meters in which the air is captured in a specific volume and each filling of this volume is indicated on a meter or recorder.

Organic membrane (cellulose derivatives, usually nitrocellulose or acetylcellulose) or nuclepore filters have the highest efficiency for collecting particles of small size but allow only relatively low flow-rates. Polystyrene filters (microsorban) allow high flow-rates but are difficult to ash, which may be necessary, e.g. for activation analysis procedures associated with radiochemical separations. The efficiency of filters decreases with decrease of particle size, increase of pore size of the filter and with increasing particle velocity at the face of the filter. Organic filters are suitable for the collection of particles as small as 0.03 μm in diameter. Glass fibre filters are used with high-volume air samplers, as they allow very high air flow-rates, but have relatively low collection efficiency [2, 3]. The data in Table 2.1 demonstrate the applicability of various filters [1].

In many methods of nuclear environmental analytical chemistry [e.g. activation analysis, XRF, PIXE (particle-induced X-ray emission)] filter impurities are a very significant factor in the choice of the filter and in the determination of the detection limits of the analytes. The content of trace elements in various filter materials is presented in Table 2.2.

2.1.2 Electrostatic precipitation

Electrostatic precipitation is used for the separation of aerosols from large volumes of gases or air, with high efficiency for small particles. A large potential difference is maintained between two electrodes, causing an intense electric field. The sampler consists of a sampler body, flow meter, suction pump and high-voltage power source (Fig. 2.3). The particles are collected on the dust-collecting electrode and may be washed or brushed out for analysis.

2.1.3 Thermal precipitation

This method is based on the principle that a thermal force (a force greater than that caused by convection) acts on a body suspended in a gas that is not in thermal equilibrium. Thus, the particles suspended in the gas migrate

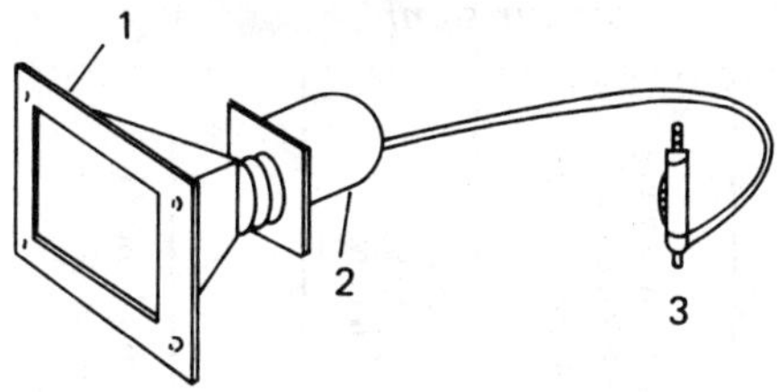

Fig. 2.1 — High-volume air sampler: 1 — filter holder, 2 — air suction assembly, 3 — flow-rate measuring assembly (indicating flow meter). (Reproduced from [1] by permission of the copyright holders, Kodansha Ltd., Tokyo).

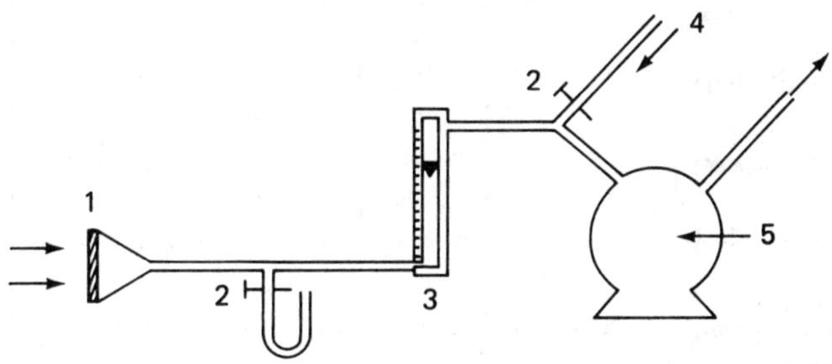

Fig. 2.2 — Low-volume air sampler: 1 — filter holder, 2 — stopcock, 3 — flow-meter, 4 — by-pass, 5 — pump. (Reproduced from [1] by permission of the copyright holders, Kodansha Ltd., Tokyo).

from a zone of high temperature to one of low temperature because of fluid creep. In general, the thermal force is negligible if the temperature gradient is less than about 750°C/cm. In the sampler the air is drawn slowly through a narrow chamber across which extends a heated wire, particulate matter being deposited upon the adjacent collecting surface. Collection is almost 100% complete for particles in the size range 0.01–10 μm. Nevertheless, the method is still not widely used as an ambient-air sampling technique. No particle shattering takes place, so this technique should be very suitable for sizing of air dust particulates [2, 4].

2.1.4 Gravitational settling

Gravitational settling is used only to collect large particles (20–50 μm) that fall out of the atmosphere under the force of gravity. It requires no vacuum source or metering system and the dust is collected on glass slides, plates, dishes, trays or in jars.

The deposit gauge of the British Standards consists of a collecting bottle (10–20 l.), hard-glass collecting funnel, bird-screening net, inverted funnel, rubber tube and metal stand (Fig 2.4). Dust jars with a polyethylene collecting container are widely used in the USA.

The gravitational settling method does not provide data on volumetric

Table 2.1 — Properties and applicability of various filters. (Reproduced from [1] by permission of the copyright holders, Kodansha Ltd., Tokyo).

Material	Filter type	Pore size (μm)	Electric charge[1]	Pressure loss[2]	Initial collecting efficiency[3]	Hygroscopicity[4]	Applicability[5] for: Low volume air sampler	High volume air sampler	Atomic-absorption spectrometry	Activation analysis	Fluorescence X-ray analysis	Remarks
Glass fibre	Toyo GB-100R		○	○	○	○	○	○	○	×	△	
	Gelman type A		○	○	○	○	○	○	○	×	△	High Zn content
	Whatman type A		○	○	○	○	○	○	○	×	△	High Zn content
	Millipore AP-20		○	○	○	○	○	○	○	×	△	High Zn content
	MSA 1106 BH		○	○	○	○	○	○	○	×	△	
Quartz fibre	Tissue Quartz 2,500 QAST		○	○	○	○	○	○	○	△	△	SiO_2 99%
	Gelman Quartz Paper type I		○	○	○	○	○	○	○	△	△	SiO_2 98.5%
	Gelman Quartz Paper type II		○	○	○	○	○	○	○	△	△	SiO_2 99.5%
Polystyrene	Microsorban (Sartorius)		○	△	○	△	○	△	○	△	×	
Poly(vinyl chloride)	Yumicron MF-100	1	○	△	○	○	○	△	○	×	○	High Cl and Ti content
Nitrocellulose	Millipore-RA	1–3	△	×	○	△	○	×	○	○	○	
	Sartorius SM	1–3	△	×	○	△	○	×	○	○	○	
	Toyo TM-100	1–3	△	×	○	△	○	×	○	○	○	
Cellulose acetate	Gelman GA-3	1–3	△	×	○	△	○	×	○	○	○	
Fluorine resin	Polyfuron PF-3 (Toyo)		○	△	○	○	○	×	△	○	—	Fibrous
	Millipore Mitex LS		○	△	○	○	○	×	△	○	—	Fibrous
	Fluoropore AF-07P (Sumitomo Denko)	10	×	×	○	○	○	△	○	○	—	Membrane type
	Millipore Florinert FA	1	△	△	○	○	○	△	○	○	—	Membrane type

[1]○, Trace; △, moderate; ×, too high for use.
[2]○, Very small; △, moderate; ×, large.
[3]○, Not less than 99%.
[4]Moisture content: ○, 0.1–0.5%; △, 1–2%.
[5]○, Good; △, sometimes good; ×, not good.

Table 2.2 — Content of trace elements in filter materials (ng/cm^2)

Metal	Polystyrene (Delbag)	Cellulose ester (Millipore 0.45 μm)	Cellulose paper (Whatman No. 41)	Glass fibre	Organic membrane	Silver membrane
Ag	2		2			
Al	20	10	12			
As				80		
Ba	500	100	100			
Be				40	0.3	200
Bi					1	
Ca	300	250	140			
Cd					5	
Co	0.2	1	0.1		0.02	
Cr	2	14	3	80	2	60
Cu	320	40	4	20	6	20
Fe	85	300	40	4000	30	300
Hg	1	1	0.5			
Mg	1500	200	80			
Mn	2	2	0.5	400	10	30
Mo					0.1	
Ni	25	50	10	80	1	100
Pb				800	8	200
Sb	1	3	0.15	30	0.1	
Sn				50	1	
Ti	70	5	10	800	2000	200
V	0.6	0.09	0.03	30	0.1	
Zn	515	20	25	160000	2	10

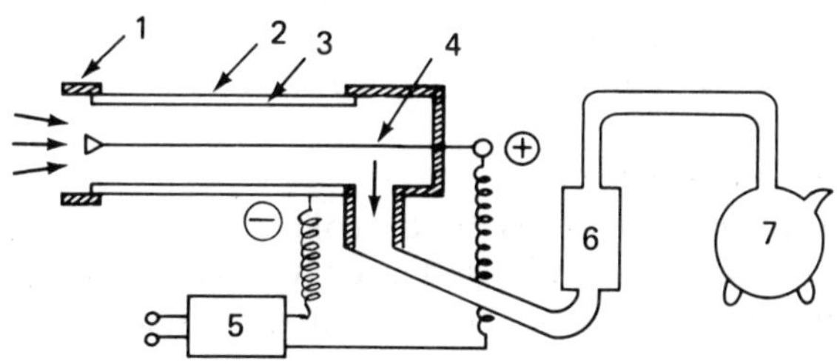

Fig. 2.3 — Electrostatic dust sampler: 1 — insulator, 2 — dust collecting electrode, 3 — dust collecting tube, 4 — discharge electrode, 5 — high-voltage power source, 6 — flow meter, 7 — suction pump. (Reproduced from [1] by permission of the copyright holders, Kodansha Ltd., Tokyo).

concentration and is beset by the complicating effects of wind velocity, direction, turbulence and rain. It is therefore, erratic, unpredictable and usually non-reproducible.

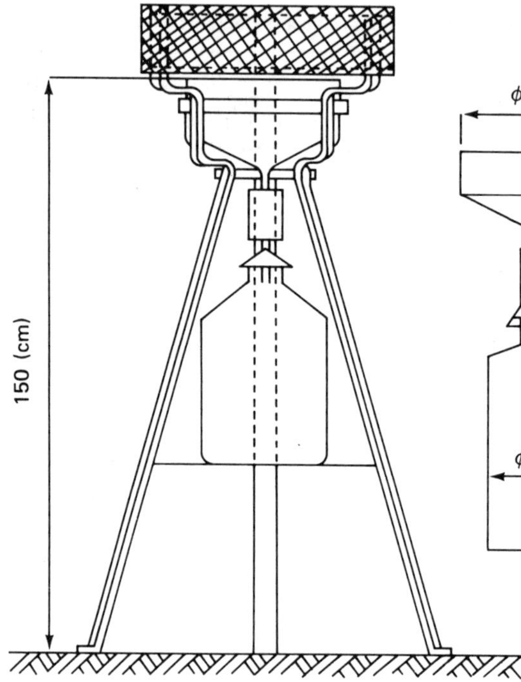

Fig. 2.4 — British Standards deposit gauge.

2.1.5 Centrifugal separation with cyclones

A cyclone collector is a structure without moving parts in which the inlet gas (air) stream passes down the inner periphery of the cyclone in a spiral and out through an exhaust tube. Centrifugal forces tend to drive the suspended particles to the wall of the cyclone. The efficiency depends on the type of cyclone used, and may reach 95–99% for 15–50 μm particles.

2.1.6 Impingement

Impingement is a process for the collection of particulate matter in which the air being sampled is directed forcibly against a surface. In *dry impingement* the particulate matter is retained on the surface against which the stream is directed: the collecting surface may be treated with a film of adhesive. *Wet impingement* is a process of impingement in a liquid which retains the particulate matter.

An *impinger* is a sampling instrument employing impingement for the collection of particulate matter. Commonly, this term is applied to two specific instruments, the midget and standard impingers, both of which use wet impingement.

The *midget impinger* uses a liquid volume of 1 ml and an air-flow of 0.1 ft^3/min. The *standard impinger* uses a liquid volume of 75 ml and an air-flow of 1 ft^3/min. The collection efficiency is nearly 100% for particles of $\geqslant 1$ μm size at the optimum sampling rate of 1 ft^3/min.

Commercial types of *dry impingers* (*impactors*) usually consist of a series of collectors for progressively smaller sizes of particles. If such a *cascade impactor* functions in an ideal manner, each stage collects all particles above a certain 'cut-off diameter'. At each successive stage the cut-off diameter would be smaller than in the previous stage (Figs. 2.5, 2.6). Impactors may

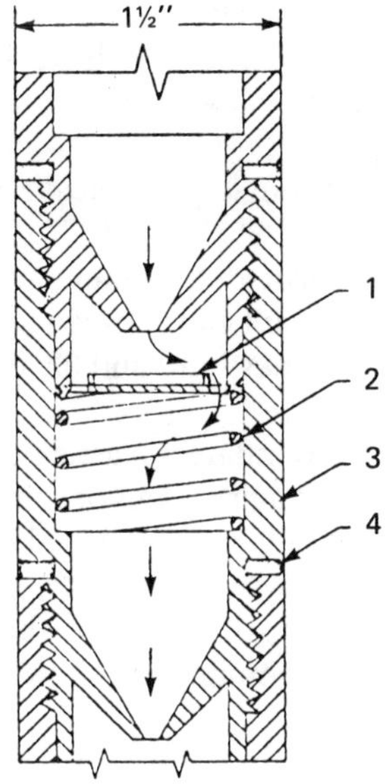

Fig. 2.5 — Stages of a typical cascade impactor (Monsanto): 1 — collection cup, 2 — spring, 3 — jet spindle, 4 — gasket.

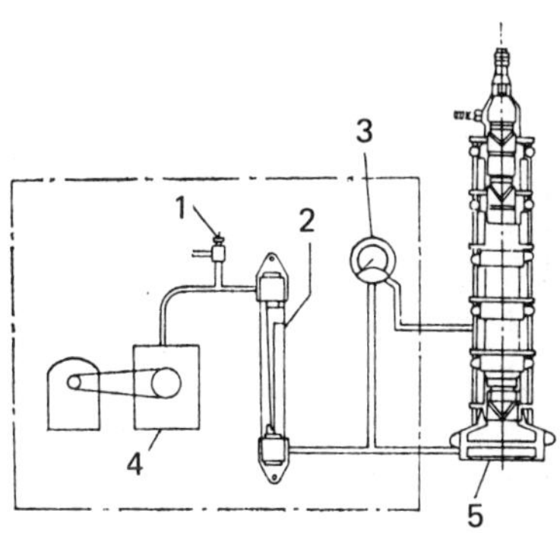

Fig. 2.6 — Diagram of a cascade impactor: 1 — leak valve for flow regulation, 2 — flowmeter, 3 — pressure gauge, 4 — vacuum pump, 5 — cascade body. (Reproduced from [1] by permission of the copyright holders, Kodansha Ltd., Tokyo).

have either a single orifice per stage or a large number of orifices per stage. In all cases the impaction surfaces, (coated metal or glass), must be soft or sticky for adequate retention of particles on impact, and to avoid bounce or dispersion of a particle deposit. Impactors may be used to measure particle sizes *in situ* in a combined sampling and sizing operation [1, 2, 4–8].

2.1.7 Collection of gaseous pollutants

The methods for collection of gaseous air pollutants (e.g. SO_2, SO_3, NO_x, O_3, H_2S, CO, HF, HCl, hydrocarbons) which utilize a wet chemical analytical approach fall within the following categories [7]:

- — air is drawn through a chemical solution which reacts with the gaseous pollutant,
- — air is passed through a suitable absorbing medium, e.g. charcoal, from which the pollutant can subsequently be recovered,
- — air is allowed to pass into and fill an inert container by either liquid or air displacement,
- — air is allowed to inflate a plastic bag — 'grab sampling',
- — air is passed through a series of collectors cooled to selected temperatures and the pollutants are condensed out in either liquid or solid form.

Different designs of impingers, bubblers and glass-frit assemblies commonly used in air sampling are shown in Fig. 2.7.

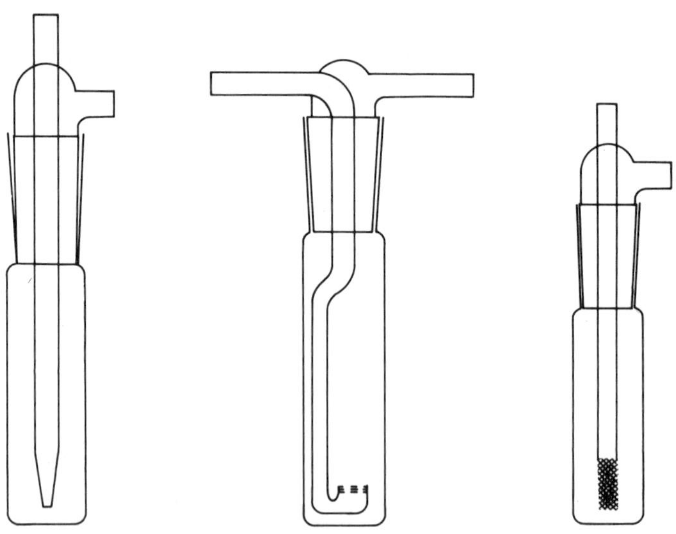

Fig. 2.7 — Different designs of impingers, bubblers and glass-frit assemblies commonly used in air sampling.

2.1.8 Air sampling of radioactive materials

An important guide to air sampling of radioactive materials is ANSI N13.1–1969 [9].

When the concentration of the radioactive component is high, the sampling procedure is simple. An enclosed metal or glass vessel equipped with inlet and outlet valves is filled simply by driving air through the vessel and subsequently closing the valves. In the case of lower concentrations of the contaminating component, it is necessary to concentrate this radioactive component during the sampling procedure, with the help of either sorption or compression. The absorption of carbon dioxide ($^{14}CO_2$) in alkali and the adsorption of radon on charcoal are examples.

Methods for sampling can be divided into two groups. In the first, the alpha, beta, or gamma activity (or gamma spectrum) is to be measured. In the second, a radiochemical analysis for certain radionuclides is to be performed.

For direct measurement, the sample is prepared by drawing a certain amount of air through a filter. To measure alpha and beta activities, the aerosols have to be retained on the filter surface. Membrane filters, plastic filters, and glass fibre filters are used in practice. For gamma activity measurement, the filters need not retain the sample at the surface.

When a radiochemical analysis has to be performed, the retention of aerosols on the filter surface is not necessary, and those filters are preferred that yield low amounts of ash. This requirement is best satisfied with cellulose filters. In addition to filtration, radioactive aerosols can also be concentrated by electrostatic or thermal precipitation [10].

Although a number of air monitoring systems are commercially available, their sensitivity is generally much lower than that of collection and laboratory analysis. In addition, their initial cost and field maintenance can be prohibitive if an adequate number of stations is to be established [11].

2.1.9 Air sampler performance criteria [12]

A suggested list of air sampler performance criteria is presented, as given in the DOE Guide [12].

(1) The system should be designed to take a representative sample of the ambient air on a continuous basis. The preferred sampling height is approximately 1.5 m above ground level, in a location free from unusual micro-meteorological or other conditions (e.g. proximity of large buildings, vehicular traffic) that could result in artificially high or low air concentrations.
(2) The linear flow-rate across particulate filters and charcoal cartridges should be maintained between 0.3 and 1.0 m/sec.
(3) A fixed sampling rate should be used, constant to within ±20% during normal operation and expected filter-loading. Total airflow or total running time should be indicated.
(4) The assembled sampling system should be leak-tested and flow-calibrated in the field.

(5) The entire system should be inspected and tested at least quarterly and recalibrated and maintained when indicated.
(6) The air-sampling system should be protected from the weather, and should be housed in a locked facility to afford a measure of security from accidental or wilful damage or tampering.

2.2 AQUATIC ENVIRONMENT SAMPLING

For a reliable sampling programme, it is necessary to take into account the goal of the programme, the nature of the system, the nature and concentration ranges of the species to be determined and the measurement techniques that will be used. Close attention must be given to the type of sample, the sampling equipment, sample containers and preservation, holding times and proper preconcentration techniques.

2.2.1 Water sampling

2.2.1.1 Types of samples

Grab or *catch samples* are samples collected at a particular time and place. Their validity depends on the temporal and spatial homogeneity of the system.

Composite samples. If the composition of a system is known to vary with time, then a series of grab samples should be taken at appropriate intervals. In favourable cases, these can be combined into a smaller number of larger samples, and an aliquot taken for analysis. Such composite samples represent time averages and lead to savings in time and money, but they must be taken with great care. They tend to smooth out temporal fluctuations, and may mask important variations in the system. Also, they are not suitable for parameters which may change quickly once the grab sample is taken, such as pH, dissolved gases or temperature. Such parameters should be measured on site, immediately after the grab sample is taken. It is usually advisable to combine individual grab samples in volumes proportional to flow-rate for streams and effluents.

Integrated samples. If a system is known to vary with location, it may be possible to combine a set of grab samples taken from different locations as simultaneously as possible. Such mixtures can be quite useful in cases where temporal variations are relatively unimportant, such as selected rivers. As with composite sampling, prior information regarding homogeneity of the system is required. The acquisition of reliable integrated samples requires special precautions so that a given grab sample is not contaminated with surrounding water (e.g. sampling at different depths).

2.2.1.2 Sampling equipment and containers

A variety of equipment is available for different sampling situations. Kemmerer and Van Dorn samplers are generally used for grab sampling of lake and river waters. Figure 2.8 includes diagrams of these two devices [13]. For sampling of sediments, dredge-type devices include the Ponar,

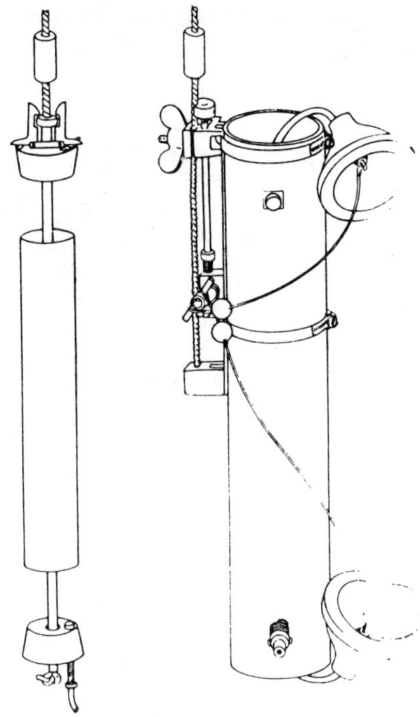

Fig. 2.8 — Structural features of common water samplers, Kemmerer (left) and Van Dorn (right).

Orange peel, Petersen, Smith–McIntyre, Ekman and Shipek samplers. Illustrations of some of these devices are given in Figs. 2.9. and 2.10 [13]. Various core-type devices are used for depth sampling, including the Phleger, Kajak–Brinkhurst and Wilding corers as shown in Figs. 2.11 and 2.12 [13].

Composite sampling of streams and effluents is frequently done with automated equipment which collects either a given fraction of the flow, or a specified number of grab samples in a fixed period. Such equipment is often custom-made and requires a good control and maintenance programme.

Ideally, sample containers are convenient to use and transport, will help prevent detrimental changes in sample composition, and will not interfere with subsequent measurements. However, there is no material suitable for all purposes. In general, polyethylene or borosilicate glass bottles are used. Metal and soft-glass containers are never used, because of possible contamination of the samples. Plastic is not used when samples are to be analysed for oil/grease, phosphorus, dissolved oxygen and most organic compounds. Glass is not used when fluoride and silica are to be determined [14].

All sample containers must be thoroughly cleaned before and after use, stored carefully, and taken to the field without becoming contaminated. At the sampling site, they should be rinsed with a few portions of sample water just prior to filling.

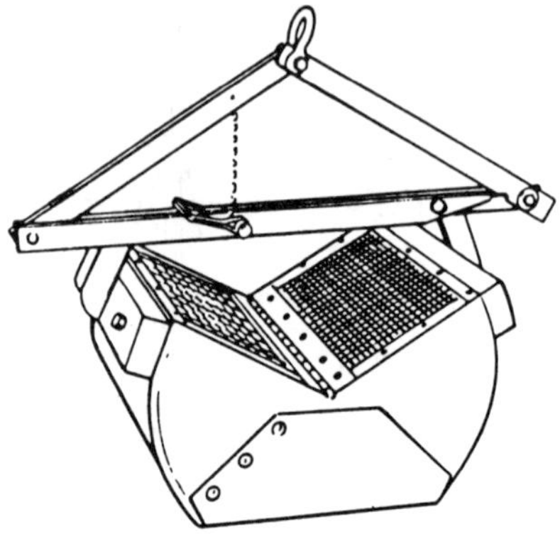

Fig. 2.9 — Ponar grab sampler.

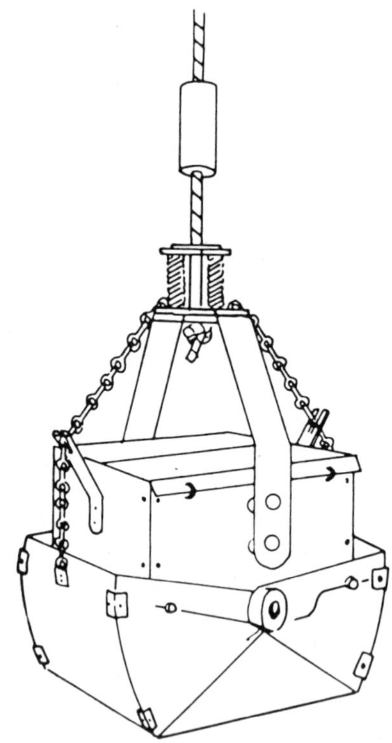

Fig. 2.10 — Ekman grab sampler.

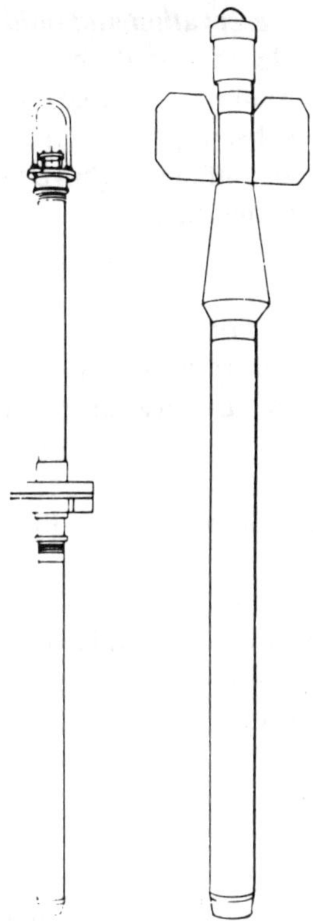

Fig. 2.11 — Phlegar core sampler.

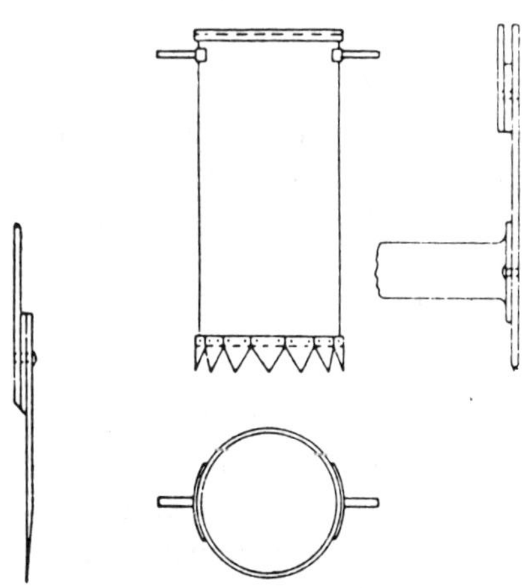

Fig. 2.12 — Wilding core sampler.

2.2.1.3 Sample preservation and holding times

The goal of sample preservation is to prevent any undesirable changes between the time samples are taken and measurements are made, without interfering with subsequent analysis. There is no perfect preservation method, and the techniques we have are few and of limited usefulness.

Undesirable changes include:

— changes in speciation,
— changes in concentrations or activities,
— movement of species between phases,

These changes can be brought about by a host of reactions, such as:

— sorption/desorption on sample containers or suspended particles,
— precipitation/dissolution,
— settling/suspension,
— gas transfer,
— redox reactions,
— reactions with ligands,
— photochemical reactions,
— biologically mediated reactions such as photosynthesis and respiration.

This is a formidable list of reactions, and many are coupled. It is almost impossible to prevent many of them completely. The best that can be done is to retard them or minimize their effects. Available preservation techniques include [13, 14]:

— refrigeration at 4°C,
— adjustment of pH with appropriate acids or bases,
— addition of biological inhibitors,
— addition of $Na_2S_2O_3$, ascorbic acid, zinc acetate in special cases.
— filtration,
— storage in the dark.

The most common method is refrigeration at 4°C. Freezing is not recommended because of possible phase separation and loss of gases.

Sample holding times [13, 14] vary considerably. Measurements should be made immediately for parameters which change quickly, such as pH, temperature, sulphite and dissolved gases. A maximum of 1–2 days is recommended for measurement of species such as bromide, cyanide, chromium(VI), nitrate, nitrite, orthophosphate, surfactants and turbidity. From a few days to some weeks is permissible for determination of many organic compounds after extraction from the water matrix, and up to 6 months is tolerable (but not desirable) for many metals.

It should be emphasized that recommendations for sample containers, preservation and holding times are specific for each parameter, and sources such as references [13] and [14] should be consulted for each analyte of interest.

The principles discussed above apply to sampling for the measurement of radionuclides. Perhaps the most significant factor is that radioactive materials are generally present in extremely low concentrations in water, and so sorption and volatilization are particularly important. It has been recommended [13] that preservatives be added at the time of sampling unless suspended and dissolved fractions are to be separated. Concentrated hydrochloric or nitric acids may be used to adjust the pH (usually to <2). Acidified samples should be held for at least 16 hr before analysis. All preservatives and reagents should be tested for radioactivity, as well as all sampling equipment and containers [15].

2.2.1.4 Sample preconcentration

Because of the extremely low concentration of many water constituents (μg/l.), samples frequently must be concentrated prior to analysis. An ideal method must produce the desired concentration range with reasonable volumes of raw water samples and within reasonable times, be convenient to use under a variety of conditions such as in the field, not change the identity of the analytes, and allow for complete recovery of the analyte.

The need for and type of preconcentation steps used in nuclear environmental analysis depend on the limit of detection, maximum amount of sample that can be handled, expected concentration levels, and interferences. Preconcentration procedures can be partially evaluated from four aspects: concentration factor, decontamination factor, recovery and specificity [16].

The *concentration factor, X,* is defined as:

$$X = C/C_0,$$

where C is the concentration in the preconcentrated aliquot and C_0 that in the original water sample.

The *decontamination factor, Y,* can be defined as

$$Y = \left[\frac{\text{analyte concn. after preconcn.}}{\text{analyte concn. before preconcn.}}\right] \Big/ \left[\frac{\text{interference concn. after preconcn.}}{\text{interference concn. before preconcn.}}\right]$$

and the *recovery, R,* is defined as:

$$R = 100\left[\frac{\text{amount of analyte collected}}{\text{amount of analyte in raw sample}}\right]\% .$$

The *specificity* for a given compound, S_i, is defined as:

$$S_i = \frac{100k_i[\mathrm{A}]_i}{\Sigma k_i[\mathrm{A}]_i},$$

where $[\mathrm{A}]_i$ is the mass of analyte i, and k_i is the corresponding specific sensitivity.

There are several techniques available for preconcentration [16,17].

Evaporation. This can be used for non-volatile materials. No decontamination is expected, there will be minimum specificity, and X values are usually between 100 and 1000. Recovery will depend on the volatility of the analyte at the temperature used.

Drying is used primarily for biological materials and to obtain samples of constant weight. There is generally no decontamination and no improvement in selectivity. Concentration factors are usually low, e.g. <1.5. Recovery can be a problem.

Freeze drying (*lyophilization*). This method minimizes loss of volatile components, but is not suitable for large volumes of sample. Concentration factors may be around 10.

Dry ashing is suitable only for analytes of low volatility. Concentration factors can be as large as 50, and recovery is good in favourable cases.

Solvent extraction is quite useful for the preconcentration of species soluble in organic phases immiscible with water. Useful examples include the concentration of metal ions after formation of non-ionic complexes with suitable chelating agents, and sampling for chlorinated hydrocarbons. In favourable cases, large values of X can be obtained, with good recovery, decontamination and specificity values.

Ion-exchange techniques are used to preconcentrate ions from relatively large volumes of water. In favourable cases, exchangers can be developed which are highly selective, and hence provide not only large concentration factors, but good decontamination and specificity [18]. Outstanding examples include the use of exchangers for multi-element activation analysis and the use of resin-loaded papers in X-ray fluorescence analysis. The use of chelating resins seems especially promising.

Co-precipitation, isotopic exchange and adsorption have also been used in limited cases. In general, a finely divided precipitate is either added to a sample, or it is formed *in situ*. In most cases, the preconcentration is actually a combined action of all three mechanisms, and hence, the term 'scavenging' is perhaps less misleading. As examples, the hydroxides of iron, aluminium and lead have been used as *in situ* scavengers for the measurement of arsenic and fall-out radionuclides.

2.2.2 Sampling of aquatic organisms

In selecting equipment and procedures for the sampling of aquatic organisms, it is necessary to consider the reason for the study and the nature of the communities to be sampled. It is convenient to divide the aquatic biosphere into the following communities: plankton, periphyton, macrophyton, macroinvertebrates and finned fish [13].

Plankton are microscopic, free-floating plants (phytoplankton) or animals (zooplankton). Because of their short life cycles, they respond quickly to environmental changes, and hence are useful as indicator organisms. Sampling locations should be as close as possible to those used for chemical sampling to allow for maximum correlation of results.

The most commonly used samplers for phytoplankton are the Kemmerer, Van Dorn, Niskin and Nansen samplers. In shallow waters the Jenkins surface mud sampler is preferred.

If live samples are to be examined, sample containers should be filled only partially, and held at ambient temperature. Formalin and Lugol's solution are commonly used preservatives.

The bottle-type samplers can also be used for smaller zooplankton, but larger species can avoid capture and must be sampled by other devices. The Juday trap or some kind of plankton net is commonly used (see Fig. 2.13 [13]).

Samples are generally preserved with 70% ethanol, 5% buffered formalin, or Lugol's solution. In the laboratory, sedimentation, membrane filtration or centrifugation are used for concentrating the field samples.

Periphyton are communities of organisms that grow on suitable surfaces such as stones, sticks or macrophytes. They are particularly sensitive to pollution and so are useful as indicator organisms. Sampling generally consists of installing a suitable substrate, allowing it to remain in place for a given time, and then recovering it. Typical natural substrates include rocks, sticks, etc. It should be remembered that quantitative recovery of periphyton from irregular surfaces is seldom possible.

The most widely used artificial substrate is the standard, plain, 25×75 mm glass microscope slide. Other materials used include clear vinyl plastic and asbestos.

A 5% solution of neutralized formalin or methiolate is used as preservative.

Macrophyton are aquatic vascular plants, mosses, liverworts, ferns and larger marine algae. There is a great variation in morphology and habitat (e.g. from tiny water meals, through cattails to giant cypress trees), so that the sampling equipment and procedures are usually specific to the goals of the study. The choice of the collection method depends on the vegetation density, type, height in the water column and nature of the sediment.

Sampling is usually done by hand from a known area, or by using special

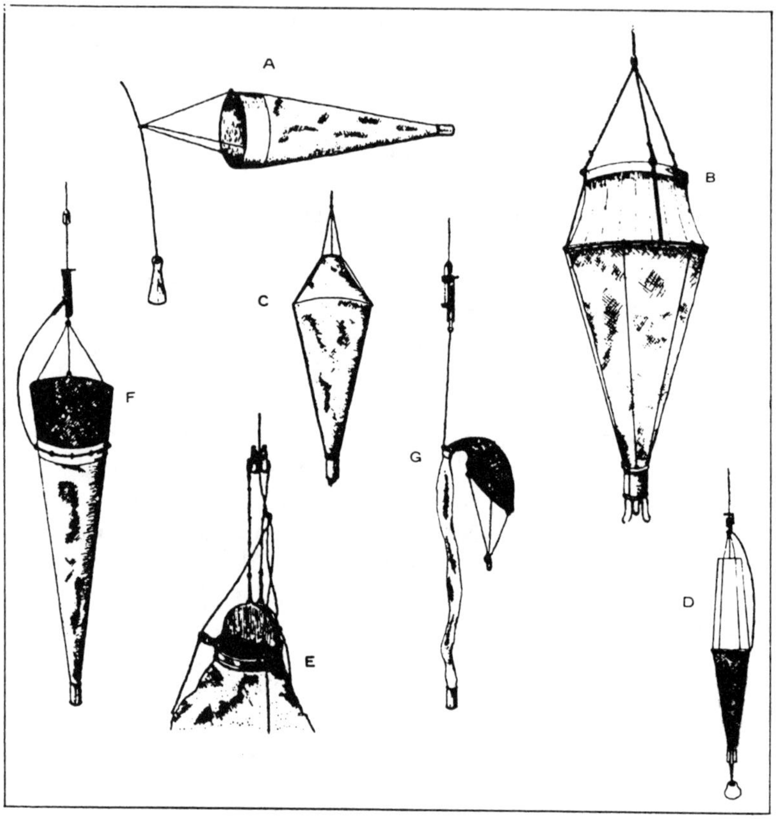

Fig. 2.13 — Plankton sampling nets. A, simple conical tow-net; B, Hensen net; C, Apstein net; D, Juday net; E, Apstein net with semicircular closing lids; f, Nansen closing net — open; G, Nansen closing net — closed.

equipment. The Ekman dredge is used for short, erect plants in soft sediments. Other samplers include a sheet metal cylinder of standard cross section (0.25 m^2), the Louisiana box sampler, the Forsberg frame sampler and the Osborne plant sampler. Figure 2.14 shows a diagram of the Osborne device [13].

Benthic microinvertebrates are animals living in the sediments. By definition, they are visible to the eye, and are retained on a U.S. Standard No. 30 sieve (0.595 mm openings). Included are insects, macrocrustaceans, molluscs, annelids, roundworms, flatworms and other species.

Sampling locations depend primarily on the goals of the study, and it is usually advisable to sample a number of locations because of possible non-homogeneity of the sample area. Quantitative and qualitative sampling devices are used. The most common quantitative grab samplers are the Ponar, Petersen and Ekman dredges. The Surber sampler is commonly used to sample riffles, although the Neils cylinder, Hess, Waters–Knapp and similar devices are also used. Depth sediment samples are obtained with

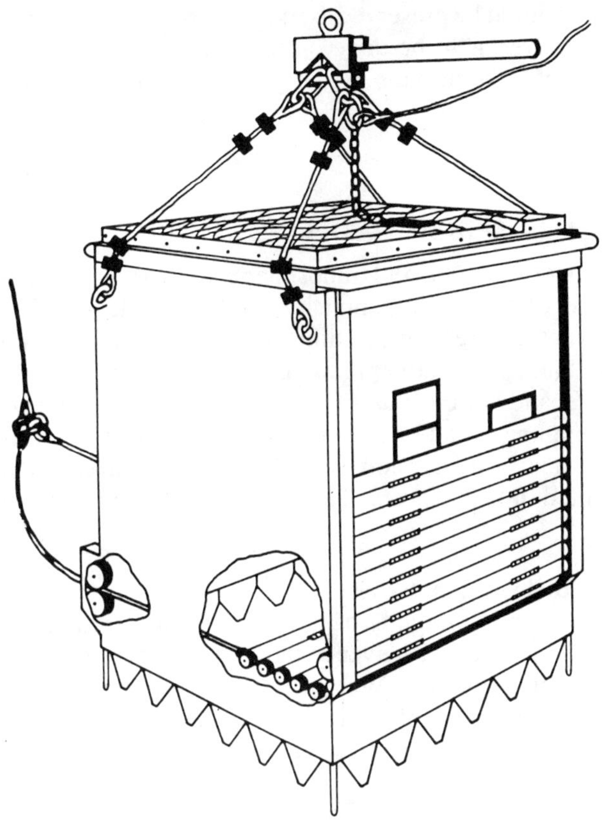

Fig. 2.14 — Osborne aquatic plant sampler.

coring devices. Drift nets are used to capture species that are moving with water currents.

Artificial substrate samplers include the modified Hester–Dendy or basket devices. Dome suction samplers are also widely used, but a diver is required to collect the samples. It is claimed that such devices give improved accuracy for site location and the opportunity to collect a relatively large number of replicate samples.

Preservatives used are 10% formalin or 70% ethanol solutions.

Fish may be collected by seining, trapping, gill or trawl netting, electrofishing, or with chemicals. Straight or bag seines are used for shallow waters. Nets are used in estuaries, lakes and rivers where fish movement is expected. Traps and weir nets are commonly used in glowing waters. Trawls are special devices used for large open waters, large rivers and marine waters.

Electrofishing devices use an electric shock to paralyse fish, followed by simple dipping to collect the specimens. The technique may be limited by depth, turbidity or conductance, and is usually more efficient at night. Rotenone is a widely used chemical for fish collection in restricted areas.

Fish should be preserved as soon as possible, preferably in the field, in ice if the analysis is to be done on the same day. For longer holding times a mixture of 10% formalin, 3 g of borax and 0.05 ml of glycerine per litre can be used.

2.3 LITHOSPHERE SAMPLING

Sampling of aquatic and atmospheric systems is relatively straightforward because of the basic simplicity of the sample matrix. The situation is somewhat different in the lithosphere, where there are far greater variations. Rocks and minerals of greatly differing accessibility, vegetation, animals, and various fluids (including water and sludges) all present different sampling situations. It is beyond the scope of this book to give a comprehensive discussion of this broad topic. Instead, a brief survey will be presented of those facets of most concern to our main topic.

2.3.1 Biological materials

Representative sampling of flora and fauna is sometimes a difficult task. Consideration must be given to the goals of the study and to the nature of the species being sampled. Why, who, how, where and when are all important questions. Because samples might be needed for the determination of trace constituents, great care must be taken to prevent losses or the inadvertent addition of analytes.

Plant materials can be sampled relatively easily, and generally without contamination problems, provided adequate attention is given to the materials used for sampling equipment (e.g. use of polyethylene gloves). Storage containers and preservatives are generally similar to those used for aquatic sampling.

The sampling of animals and human tissue is more difficult, and contamination by trace materials is a serious problem. For example, surgical steel tools cannot be used because of possible contamination with Cr, Fe, Co, Ni, Mn, Cu, Mo, Cd and Zn, or in some cases Ag, Sb, Sn and Ta [19]. The problem can be minimized by the use of tools made of silicon, plastics, Pt alloys or Ti, because these elements are usually not determined.

The analysis of samples of human origin (fluids, soft and bony tissues, hair, etc.) provides information about the presence of toxic substances. Such samples are obtained by current clinical methods [20].

The selection of foodstuffs depends on the behaviour of analytes in the food chain and the eating habits of local populations. Materials frequently analysed for radioactive contamination include milk, cheese, fruits, vegetables and honey. For milk, no unique sampling methods are required, but care should be taken that samples are not diluted with products from different areas. Preservatives should be added to prevent souring or curdling. Refrigeration can be used for shorter holding times (a few days). Polyethylene containers are recommended.

2.3.2 Soils

The location, number and type of sample sites depend on the local topography, purpose of the study, analytes and proximity to sources of contamination. Because soil characteristics can change significantly over relatively short distances, it is sometimes advisable to sample randomly from a grid. The number of samples and sample density (samples per unit area) depends on the homogeneity of the matrix and purpose of the study. It is generally appropriate to use composite samples for better statistics and greater cost-effectiveness (e.g. one composite from 10–30 grab samples).

The interpretation of data obtained should take into account factors such as relief, soil type, land use, chemical additives such as fertilizers and pesticides, and type of ground cover.

Sampling equipment includes shovels and trowels for surface samples and various coring devices for depth samples, such as those shown in Fig. 2.15 [21]. A depth of 20 cm is frequently used, and the top 5 cm is

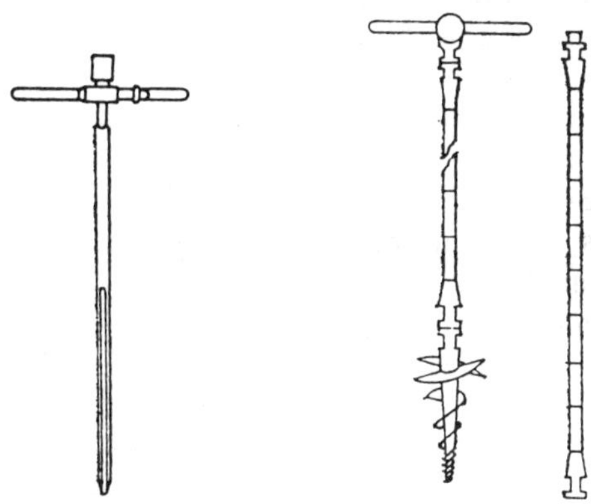

a b

Fig. 2.15 — Corer for soil sampling.

generally handled separately. Samples are transported and stored in plastic bags.

In the laboratory, composite samples are typically quartered and air-dried. Rocks and plant materials are removed. After drying, the samples are ground and sieved with a 2-mm mesh screen and stored [22].

2.3.3 Sludges

Depending on the goals of the study and nature of the sample matrix, single or composite samples may be taken. If the sludge has a high liquid content and is homogeneous, standard water samplers are used. For materials of higher solids content, a Rawson sampler, which is essentially a suction device, can be used (Fig. 3.16) [23]. Suction is provided by a rubber ball

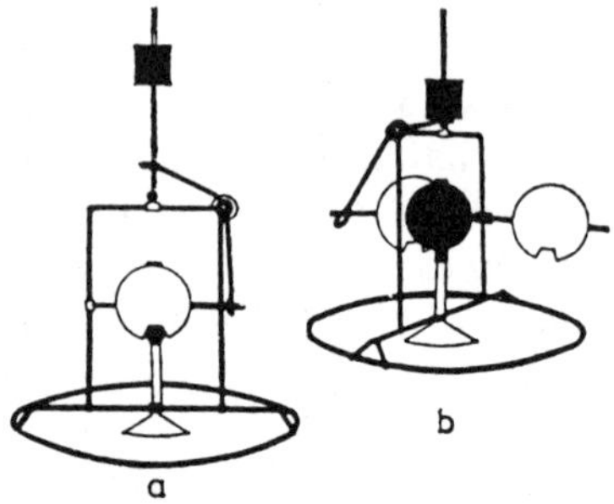

Fig. 2.16 — Rawson sampler.

which is squeezed above the surface with metal plates and allowed to open when in contact with the sludge deposit. The captured material is brought to the surface and removed from the ball.

Semi-solid sludges are generally sampled with the same dredges used for sediment sampling, including the Mason dredge, shown in Fig. 2.17 [23].

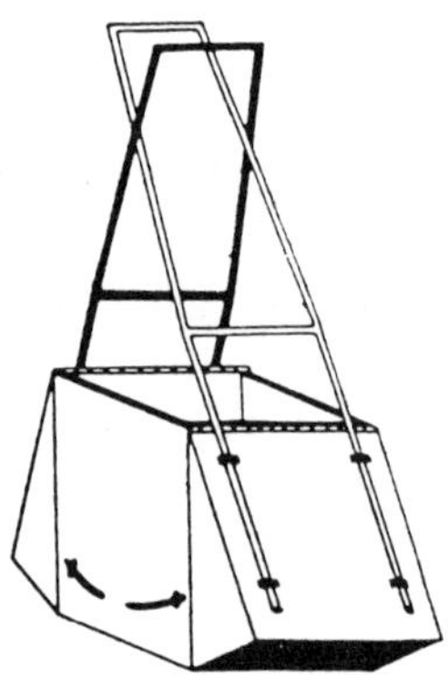

Fig. 2.17 — Mason sampler.

Plunger type samplers are also used. Examples of such devices are shown in Fig. 2.18 [24].

2.4 ADDITIONAL COMMENTS ON SAMPLING THE ENVIRONMENT FOR THE MEASUREMENT OF RADIOACTIVE CONTAMINATION

Special consideration must be given to sampling for the measurement of radioactive contamination in the vicinity of nuclear facilities (all phases of the nuclear fuel cycle, radioisotope production, accelerators, medical and

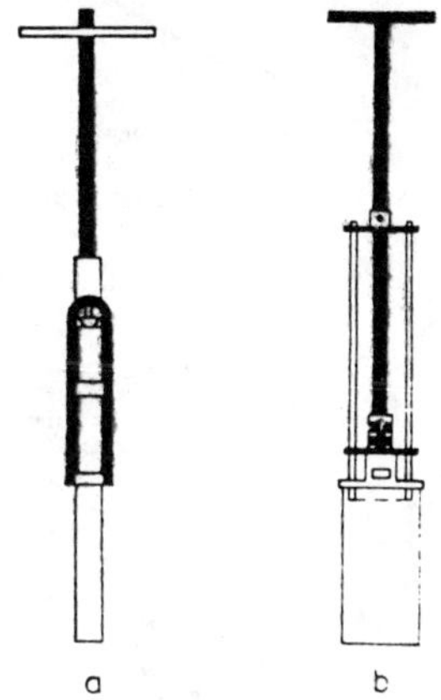

Fig. 2.18 — Lenz sampler.

research facilities), or in the event of accidents involving the release of radioactive materials during shipment or storage (e.g. spills during shipment). In such cases, a surveillance programme is needed to provide information used for dose calculations, comparison with regulatory standards, assessment of environmental contamination, counter-measures and public information.

The objectives of such a programme are [11] as follows.

(1) Assessment of actual or potential exposure to humans.
(2) Evaluation of long-term trends in the environment.
(3) Study of the fate of contaminants.
(4) Baseline information and database maintenance.
(5) Evaluation of off-site sources of radiation.
(6) Compliance with regulations.

Factors which need to be considered include the following:

(1) Potential hazards, including quantities and radiotoxicities.
(2) Size and distribution of the exposed population.
(3) Cost-effectiveness.
(4) Availability of resources, including personnel, equipment, techniques, time and money.

It must be possible to make adequate comparison with standards, reference materials and background measurements.

The flow diagram given in Fig. 2.19 can be used as an aid in relating data requirements and critical path analysis to overall programme planning [11].

In the case of nuclear facilities, both off-site and on-site sampling is

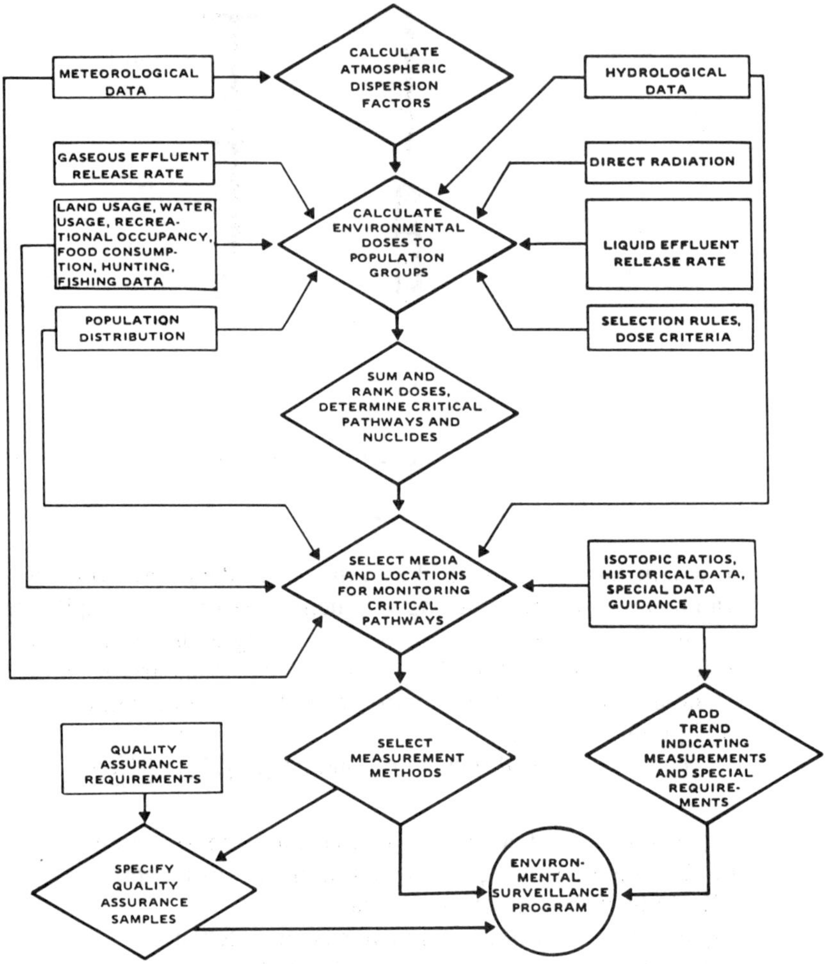

Fig. 2.19 — Environmental surveillance programme design process [11]. (Reproduced by permission of the copyright holders, CRC Press, Inc., Boca Raton, Florida).

required, in conjunction with pre-operational and operational programmes. A pre-operational programme should consist of:

(1) identification of critical exposure pathways,
(2) identification of population groups susceptible to exposure in the event of a release from the facility,
(3) selection of types of samples and sampling sites,
(4) accumulation of a background database,
(5) interpretation of information obtained.

Once the facility is operational, continuous surveillance is needed to monitor effluents and the surrounding environment. An example of a sampling programme has been provided by the U.S. Environmental Protection Agency [25]. Such programmes rely heavily on the critical exposure pathway concept. The U.S. Department of Energy (DOE) has also provided guidelines. A summary of major items is given in Table 2.3 [12]. Good sources of information regarding the entire question of radiation surveillance are the HASL document [24] and the DOE Guide [12].

Table 2.4 [10] indicates the important radionuclides in evaluating contamination of the environment in the case of release from a nuclear facility,

Table 2.3 — General guide for environmental surveillance programmes. (Reproduced from [11] by permission of the copyright holders, CRC Press Inc., Boca Raton, Florida).

1. Every exposure pathway should be routinely measured so that, as determined by site-specific exposure pathway analysis and either predicted or historical release rates, it may contribute more than 10% of a total environmental dose of site origin, if the total exceeds:
   a. 1 mrem/year to the whole body or specific organ dose of individuals or critical population groups; or
   b. 100 man-rem/year whole body dose per 1 million population within an 80 km (50 mile) radius of the release point (or centroid of a cluster of release points).
2. For dose calculation, actual measurements need not be made on more than two media for each critical radionuclide/pathway combination, one of which may be the effluent stream.
3. If reasonably achievable, routine measurement techniques for dose calculation should have minimum detections limits no higher than those equivalent to the dose criteria given in item 1.
4. 'Background' or 'control' location measurements should be taken for every critical nuclide/pathway combination for which environmental measurements are used in the dose calculation.
5. Dose calculations from direct environmental measurements should be based only on statistically significant differences between the point of measurement and background or control data. For this purpose, statistical comparisons should be made at the 95% confidence level (a difference in mean values for the assumed statistical distributions of at least two times the pooled standard deviations of the data).
6. Gross activity analyses as such should be used only as trend indicators, with supporting analyses to relate results to specific nuclide concentrations or doses.
7. The exposure pathway analysis should be documented, with data references and local assumptions clearly distinguished, and with critical nuclides, pathways, and population groups described.
8. Sampling or measurement intervals for each critical nuclide/medium combination should not exceed twice the half-life of the nuclide to be measured.
9. Where significant periodic variations in environmental concentrations may be expected, samples or measurements should be either continuous or at an interval less than half the expected peak-to-peak interval.

Table 2.4 — Samples for evaluating the human environment contamination (Reproduced by permission of the copyright holders, Veda, Bratislava.)

Samples	Radionuclides
Air	All radionuclides dispersed with air
Water	All radionuclides dispersed with water
Soil	^{90}Sr, ^{226}Ra, ^{137}Cs
Plants	^{131}I, ^{89}Sr, ^{90}Sr, ^{137}Cs, ^{226}Ra, ^{140}Ba
Milk	^{89}Sr, ^{90}Sr, ^{137}Cs, ^{131}I, ^{140}Ba, ^{226}Ra
Milk products	^{89}Sr, ^{90}Sr, ^{137}Cs, ^{226}Ra
Meat	^{137}Cs
Thyroid gland	^{131}I
Bones	^{89}Sr, $_{90}Sr$, ^{226}Ra
Water organisms	^{89}Sr, ^{90}Sr, ^{137}Cs, ^{106}Ru, ^{141}Ce, ^{144}Ce

and Table 2.5 [10] lists those nuclides commonly measured in samples of human origin.

A somewhat different presentation of important nuclides is given in Table 2.6, taken from the work of Denham [11].

In general, sampling techniques for the evaluation of radioactive contamination of the environment in the event of a nuclear accident are based on standard procedures and are similar to those already discussed [24]. It must be remembered that the sampling programme is at least as important as the measurements made on the samples, and this is especially true for nuclear accidents.

Table 2.5 — Samples of human origin generally used in measuring radioactive contamination (Reproduced by permission of the copyright holders, Veda, Bratislava).

Radionuclide	Basic sample	Other samples	Remarks
^{3}H	Urine	—	
^{89}Sr	Urine	Faeces	
^{90}Sr	Urine	Faeces, bones	
^{131}I	Thyroid gland	Urine	
^{137}Cs — ^{137}Ba	Urine	Blood	
^{140}Ba — ^{140}La	Urine	—	
^{226}Ra	Expired air, urine	Bones, faeces	Rn is determined in expired air
^{239}Pu	Urine	Faeces	
^{210}Po	Urine, blood	Faeces	
Th (natural)	Urine	—	
U	Urine	—	

Table 2.6 — Relative importance of environmental media as a function of effluent pathway and radionuclide(s). (Reproduced from [11] by permission of the copyright holders, CRC Press, Inc., Boca Raton, Florida).

Effluent Pathway — Atmosphere

	Tritium	Noble gases	Iodine — 131	MF & AP[a]	TRU[b]
1	Air	Direct radiation	Milk	Air	Air
	Vegetables	Air	Vegetables	Vegetables	
2	Honey		Air	Milk	Deposition
	Native grasses		Animal thyroids	Direct radiation	Terrestrial biota
	Precipitation		Forage	Deposition	Soil
3				Forage	

Effluent Pathway — Hydrosphere

	Tritium	Iodine — 131	MF & AP[a]	TRU[b]
1	Drinking water	Drinking water	Fish, shellfish	Sediment
	Surface water	Milk		Fish
2	Vegetables	Surface water	Waterfowl	Surface water
	Groundwater	Vegetables	Surface water	Vegetation
	Honey	Forage	Milk	
3			Sediment	
			Drinking and ground water	
			Terrestrial biota	

[a] Mixed fission and activation products.
[b] Transuranic nuclides, primarily ^{239}Pu.

REFERENCES

[1] K. Oikawa, *Trace Analysis of Atmospheric Samples*, Kodansha, Tokyo; Wiley, New York, 1977.
[2] A. Alian and B. Sansoni, *J. Radioanal. Nucl. Chem.*, 1985, **89**, 191.
[3] A. L. Choen, *Environ. Sci. Technol.*, 1973, **7**, 60.
[4] H. F. Lund (ed.), *Industrial Pollution Control Handbook*, McGraw-Hill, New York, 1971.
[5] W. F. Merritt, *Proc. Symp. Measurement, Detection and Control of Environmental Pollutants*, IAEA, Vienna, 1976, p. 75.
[6] J. A. Robbins, *Precipitation Scavenging* (1970), CONF-700601, USAEC, Dec. 1970.
[7] J. D. Butler, *Air Pollution Chemistry*, Academic Press, London, 1979.
[8] A. L. Linch, *Evaluation of Ambient Air Quality by Personal Monitoring*, CRC Press, Boca Raton, Florida, 1981.
[9] *Guide to Sampling Airborne Radioactive Materials in Nuclear Facilities*, ANSI N13.1, American Standards Institute, New York, 1969.
[10] J. Tölgyessy and Š. Varga, *Nuclear Analytical Chemistry*, Vol. 3, *Radiochemical and Activation Analysis*, Veda, Bratislava; University Park Press, Baltimore, 1974.
[11] D. H. Denham, in *CRC Handbook of Environmental Radiation*, A. W. Klement (ed.). pp. 117–127. CRC Press, Boca Raton, Florida, 1982.
[12] J. P. Corley, D. H. Denham, R. E. Jacquish, D. E. Michels, A. R. Olsen and D. S. Waite, *A Guide for Environmental Radiological Surveillance at U.S. Department of Energy Installations,* DOE/EP-0023, Office of Operational Safety, U.S. Department of Energy, Washington, D.C., July 1981.
[13] *Standard Methods for the Examination of Water and Wastewater*, 16th Ed., American Public Health Association, Washington, D.C., 1985.
[14] *U.S. Federal Register*, 40 CFR, Part 136, Vol. **48**, No. 209, October 26, 1984. *Guidelines Establishing Test Procedures for the Analysis of Pollutants under the Clean Water Act.*
[15] U.S. Geological Survey, *Methods for Determination of Radioactive Substances in Water and Fluvial Sediments*, U.S. Govt. Printing Office, Washington, D.C., 1977.
[16] H. A. Das, A. Faanhof and H. A. Van der Sloot, *Environmental Radioanalysis*, Elsevier, Amsterdam, 1983.
[17] S. E. Manahan, *Environmental Chemistry*, 4th Ed., Willard Grant Press, Boston, 1984.
[18] J. Albaiges (ed.), *Analytical Techniques in Environmental Chemistry*, Pergamon Press, Oxford, 1982.
[19] S. R. Koirtyohan and C. A. Hopkins, *Analyst*, 1976, **101**, 870.
[20] J. Kučera and L. Obrusnik, *Rádioaktivata a životne prostredie*, 1979, **2**, s.1, č.4.
[21] M. Kutilek, *Vodohospodářská pedologie*, SNTL, Prague, 1978.
[22] J. Tölgyessy, E. Havránek and E. Dejmková, *Rádionuklidová röntgenofluorescenčná analýza zložiek životného prostredia*, Alfa, Bratislava, Czechoslovakia, 1983.
[23] M. Sedláček, *et al., Metody rozboru kalu a pevných odpadů*, SZN, Prague, 1978.
[24] J. H. Harley, *HASL Procedures Manual*, HASL-300, Health and Safety Laboratory, Atomic Energy Commission, (revised annually by the Environmental Measurements Laboratory, U.S. Department of Energy).
[25] Office of Radiation Programs, *Environmental Radioactivity Surveillance Guide*, ORP/SID-72-2, U.S. Environmental Protection Agency, Washington, D.C., 1972.

3

Preparation of standards and sample pretreatment

In many cases direct measurement cannot be made on environmental samples. Very often they must be preserved and analytes must be separated, concentrated or sometimes transferred into another phase. Generally, therefore, a number of steps may be required prior to final measurements.

Standards and reference materials are exceedingly important for the verification of each step of the total procedure and calibration of the final measurement technique. Careful attention must be given to their preparation and use.

3.1 REFERENCE MATERIALS AND PREPARATION OF STANDARDS

Generally, in order to judge the reliability of an analytical method or the results of a determination, special samples are needed that have a known composition and are as similar as possible to the samples which are to be analysed. Such samples are usually referred to as reference samples (reference materials, standard materials, standards). They are used for the calibration of instruments, verification of instrument performance, chemical yield determinations and quality control, as well as for evaluation and comparison of analytical methods.

In practice, they are substances for which one or more chemical, physical, biological or technical properties have been measured reliably and are therefore considered to be known. For chemical analysis, such materials must be homogeneous and as stable as possible. Components of interest must have been determined in a sufficient number of replicates, by commonly accepted methods and preferably at several laboratories, and the average value (after statistical treatment) is quoted. For example, certification of some reference materials at the U.S. National Bureau of Standards is done in one of three ways [1].

(1) Measurement by a method of known accuracy, by at least two analysts, or at least in two different laboratories.
(2) Measurement by two or more independent and reliable methods, which give errors that are small compared to that required for certification.
(3) Measurement with the co-operation of a group of qualified laboratories using a number of methods.

All reference materials have a certificate which includes the nature and amount of various components, methods used, and a statement of reliability. Many countries have institutions responsible for providing such materials; for example the US NBS and Glavnaya plata mer i vesov in the Soviet Union. A list of materials provided by the US NBS is given in Table 3.1 [1]. Materials available from the International Atomic Energy Agency are given in Table 3.3 [2].

Another good example of a reference material is Bowen's Standard Kale (prepared by H. J. M. Bowen of England), used by many laboratories for the analysis of plant materials. Professor Bowen prepared 91 kg of dry kale powder and distributed portions to several participating laboratories for analysis. On the basis of data received, he calculated concentration values and ranges for a group of macroelements (Na, K, Ca, Mo, O, P, S, N, Cl) as well as 30 trace elements [3–5]. An example of a set of values for the material is given in Table 3.3 [6].

It is often necessary for analysts to prepare their own laboratory standards for quality control, calibration of instruments, or analytical calculations. These are usually prepared by either static or dynamic methods. The basic principle of the *static method* is to add a defined amount of standard material (primary standard) to a known volume of water or other solvent, or carrier gas, or place it on a solid carrier (e.g. a membrane or paper filter).

In the preparation of these laboratory standards, attention must be paid to the purity and stability of reagents used, contributions from carrier materials (solvents, gases, filter media) and the calibration of equipment such as pipettes. Only reagents of the highest purity are used, and these are carefully dried in a desiccator or at a higher temperature in a dry box. Substances which can react with gases in the atmosphere (CO_2, O_2, water vapour) should be stored securely, for example in weighing bottles in a desiccator. In handling of reagents attempts should be made to minimize contamination and contact with the atmosphere, and to carry out the operation as quickly as possible consistent with good laboratory practice. Carefully distilled or demineralized water is used to make up aqueous solutions, and its purity should be monitored regularly as part of the laboratory control programme.

Solutions of working concentrations should be made by dilution of standardized stock solutions immediately before use. The stock solutions should be prepared on a regular basis with due attention to shelf-life, and stored properly. Dilution must be made with appropriate calibrated equip-

Table 3.1 — Reference materials available from the US National Bureau of Standards

Standard number	Type	Analytes, concentration
	Air standards	
1658–1659	Methane in air	1, 10 ppm
1660	Methane and propane in air	4 ppm CH_4; 1 ppm C_3H_8
1665–1669	Propane in air	2.8, 9.5, 48, 95, 475 ppm
2611–2614	Carbon monoxide in air	1, 10, 18, 45 ppm
2676	Metals on a filter	Cd, Pb, Mn, Zn
2677	Silica dust on a filter	SiO_2
1648	City aerosols	macro and trace elements, organic components
	Water standards	
1641	Mercury in water	1.49 μg/ml
1643	Trace elements in water	20 trace elements
	Biological materials	
1571	Orchard leaves	Pb, Hg and 17 other elements
1577	Bovine liver	Pb, Hg and 10 other elements
1570	Spinach	K, Ca, P and 12 trace elements
1575	Pine needles	K, Ca, P, Al, Hg and 12 trace elements
1573	Tomato leaves	K, Ca, P, Pb and 11 trace elements
	Solid materials	
1579	Pb dye dusts	Pb (11.87%)
1631	Coal	S (3 concentrations)
1632	Coal	14 trace elements
1623	Fly ash	9 trace elements
1645	River sediment	macro and trace elements
4350	River sediment	10 radionuclides

ment, and is best done in steps to avoid possible errors caused by large dilution factors. For example, if the stock solution contains 0.1 mg/ml of the analyte, and it is desired to prepare 100 ml of working standard at a concentration of 1 μg/ml, it is best to use two dilution steps: 10 ml of stock diluted to 100 ml and then 10 ml of this intermediate solution (10 μg/ml) diluted to 100 ml of working solution at 1 μg/ml.

An example of the procedure used for the preparation of laboratory standards of toxic metals (Pb, Cd, Zn, Sn, Mn, Be) on a cellulose acetate filter (MF-HA; 47 mm diameter; 0.45 μm porosity) is described in [7]. The working standard solutions were prepared by dissolution of known amounts of the pure metals in dilute, high-purity nitric acid (Sn was dissolved in

Table 3.2 — Standard reference materials from the IAEA

Code number	Type	Analytes*
V-SMOW	Water	$^{18}O/^{16}O$, $^{2}H/^{1}H$
SLAP	Water	$^{18}O/^{16}O$, $^{2}H/^{1}H$
W-4	Fresh water	17 trace elements
Air-3/1	Deposition on filter	13 trace elements
Soil-6	Soil	^{90}Sr, ^{137}Cs, ^{226}Ra, ^{239}Pu
Soil-7	Soil	28 trace elements
SL-1	Lake sediment	28 trace elements
SL-2	Lake sediment	^{137}Cs, ^{226}Ra, ^{40}K
A-11	Milk powder	12 trace elements
A-14	Milk powder	^{90}Sr, ^{137}Cs, Na, K, Ca
A-12	Animal bone	^{226}Ra, ^{90}Sr
A-13	Dried animal blood	10 trace elements
V-8	Rye flour	11 trace elements
V-10	Hay (powder)	27 trace elements
H-4	Animal muscle	14 trace elements
H-5	Animal bone	12 trace elements
H-9	Mixed human diet	multielement
IAEA-083	Simulated air filter	^{60}Co, ^{90}Sr, ^{133}Ba, ^{137}Cs, ^{210}Pb

*Non-certified information is also given for several materials

Table 3.3 — Values of the 'grand mean' for Bowen's kale sample (μg/g). (Reproduced from [6] by permission of the copyright holders, Akadémiai Kiadó, Budapest).

Element	Bowen^{A} [3] (different methods)	Bowen^{B} [4] (different methods)	Bowen^{C} [5] (neutron activation)
Al	7.35–80.1	38.2 (6.4–88.4)	37.46±7.84 (7)
As	0.127–1.80	0.141 (0.11–0.22)	0.141±0.031 (18)
Ca	41 400±2230 (41)	40850	40409±2544 (11)
Cd	1.0±0.1 (4)	0.80 (0.38–1.06)	0.76±0.192 (11)
Cl	3330±1060 (21)	3415 (2180–4450)	3711±368 (9)
Co	0.0562±0.0077 (22)	0.0581 (0.041–0.081)	0.0592±0.0103 (15)
Cr	0.331±0.155 (13)	0.308 (0.18–0.42)	0.356–0.127 (10)
Cu	4.81±0.735 (88)	4.99 (3.6–6.5)	4.679±0.644 (16)
Fe	119.5±19.5 (79)	118.3 (88–157)	117.3±16.2 (12)
K	24630±1218 (53)	24615 (20600–29300)	24248±1390 (5) INAA 20976±3701 (10)
Mg	1604±119 (43)	1572 (1350–1700)	1514±88 (6)
Mn	14.9±1.8 (83)	14.73 (12.6–18)	14.58±1.26 (21)
Mo	2.33±0.47 (44)	2.28 (1.5–3.1)	2.325±0.507 (12)
Na	2594±617 (52)	2506 (1220–3250)	2257±258 (12)
Rb	52.8±6.25	52.2 (49–57)	53.38±3.87 (7)
Sb	0.0653±0.0125 (6)	0.0689 (0.05–0.11)	0.0719±0.0173 (13)
Sc	0.00835±0.00074 (4)	0.00829	0.00779±0.00092 (7)
Se	0.148±0.0137 (20)	0.121? (0.02–0.15)	0.1375±0.017 (10)
Sn	0.160±0.037 (4)	0.26 (0.16–0.36)	0.23±0.071 (7)
V	—	0.36 (0.33–0.41)	0.366±0.03 (5)
W	0.0605±0.00123 (8)	0.0605	0.06±0.0029 (3)
Zn	31.88±4.82 (77)	33.2 (30–38)	31.85±2.09 (23)

^{A}Range or mean ± standard deviation (number of analyses).
^{B}Mean and range.
^{C}Mean ± standard deviation (number of laboratory averages).

hydrochloric acid). Small volumes (0.1 ml) of the standard solutions were placed on the filters. The amounts of the metals are given in Table 3.4. They represent an assumed human exposure level, and a given sample collection system.

In *dynamic methods* for the preparation of laboratory standards, continuous flows of fluid (water or non-aqueous solutions, gas mixtures) of known composition are diluted with pure carrier at a controlled dilution ratio. Gaseous systems are more difficult to operate and four approaches are used.

(1) Carrier gas is bubbled through a liquid solution containing the material of interest at known concentration and temperature. The resulting concentrations in the laboratory standards are calculated from the vapour pressure of the standard material at the liquid concentration and temperature used, and the flow-rate of the carrier (dilution) gas.
(2) Diffusion of the analyte gas into a flow of carrier gas under controlled conditions.
(3) Permeation of analyte gas into the flow of carrier gas through a membrane (e.g. FEP Teflon).
(4) Direct, continuous injection of known amounts of analyte gas into the flow of carrier gas (simple dilution).

3.2 PREPARATION OF AIR SAMPLES

In some cases, measurements can be made directly on air samples. Examples include the use of X-ray fluorescence (XRF) or activation analysis for direct measurement of metals in aerosols captured on filter media. Collection using surface-active materials, electrostatic precipitators or cascade impactors can also give samples which might be used for direct measurement [7].

In the case of air sampling by absorption in a collecting-liquid solution, it is often advantageous to extract the absorbed analytes or concentrate them on ion-exchange materials, followed by elution.

For trace element analysis, it is possible to dissolve the aerosols off the collecting filters and concentrate the analytes by ion-exchange or extraction. Sometimes it is necessary to combine several filters to obtain sufficient analyte for measurement.

Wet or dry ashing techniques may be needed to prepare particulates for dissolution. Aerosols and filter media are both mineralized in this case. For *dry ashing*, it is recommended to use temperatures below 500–550°C because of possible loss of volatile components (As, Se, Pb, Sn, Hg). In some cases it is preferable to use a high-frequency electric field with a flow of oxygen at reduced temperatures (30–150°C).

Wet ashing is carried out at lower temperatures, so loss of volatiles is minimized. Various oxidizing agents are used: strong acids (H_2SO_4, HNO_3, $HClO_4$), hydrogen peroxide, or mixtures of both (H_2SO_4+HNO_3+H_2O_2, or HNO_3+$HClO_4$, or H_2SO_4+H_2O_2). The efficiency of wet oxidation can

Table 3.4 — Recommended values (μg/filter) for toxic-metals laboratory standards made on filters [7]

Number		Pb	Cd	Zn	Sn	Mn	Be
1	1A	7.4	4.9	99	148		
	1B	14.8	9.9	198	297		
	1C	29.7	19.8	396	594		
2	2A	7.4	—	111	—	248	
	2B	14.8	—	226	—	295	
	2C	29.7	—	452	—	990	
3	3A	—	—	—	—	—	0.099
	3B	—	—	—	—	—	0.198
	3C	—	—	—	—	—	0.396

be improved by using a combination of higher temperatures and proper choice of mixtures of oxidizing agents. Ashing of filter papers at 500°C with or without prior wetting with sulphuric acid gives a recovery of metals from air particulates comparable to that from wet ashing with HNO_3+$HClO_4$ or low-temperature ashing. The use of glass fibre filters or silica, glass, or glazed porcelain crucibles leads to low recovery of metals [8,9].

Most nuclear analytical methods (e.g. XRF and activation analysis) can be used only for the determination of elements. It is usually impossible to use them for the determination of speciation, chemical bonds, or physical properties such as size distribution of particles. For this reason, a combination of methods is needed for a more thorough study of aerosols.

3.3 PREPARATION OF WATER SAMPLES

In some cases, it is sufficient to homogenize the sample by simple mixing. If it is desired to analyse separate phases (e.g. dissolved and suspended), or if the analyte may be present in more than one phase, pretreatment is required.

Particulates or insoluble forms of the analyte are usually separated by centrifugation or filtration. Centrifugation is used for larger amounts of suspended material. Filtration media are chosen according to the chemical/physical nature of the analyte; cellulose acetate membrane or glass fibre filters are commonly used.

For trace element determination, separation and preconcentration steps can be combined in favourable cases, as for example liquid–liquid extraction, sorption with activated charcoal, scavenging with hydrous oxides (Fe, Al), electroplating, ion-exchange or some combination of these. Extraction and ion-exchange are the most commonly used.

A more recent development [10,11] is the use of foamed plastics, especially open-cell type resilient polyurethanes, for separation and/or preconcentration of trace analytes from environmental samples. The main advantages of these materials are the quasi-membrane structure which allows rapid solid/liquid or solid/gas contact in column operation, and the ability to immobilize a variety of reagents as collectors and so act as a highly versatile solid-phase support. The resilience of these foams permits their use in special modes such as automated batch squeezing or pulsating column operation.

The foams can be used in the field for both air and water sampling. Moreover, in favourable cases, they can be used equally well for either direct or indirect multielement determinations.

3.4 PREPARATION OF SAMPLES FROM BIOLOGICAL MATERIALS

The time between sampling of biological materials and analysis should be kept as short as possible. For storage times of a few (2 or 3) days, refrigeration at 4°C may be adequate. In favourable cases, holding times can be prolonged by the addition of chemical preservatives (e.g. formalin, alcohols), but care must be taken that these additives do not interfere with subsequent analysis. Radiation preservation and dehydration are also used. Dehydration can be done by drying, combustion and lyophilization. *Combustion* can cause a large reduction in sample volume, which may be desirable, but it cannot be used in the determination of elements which form volatile species (Hg, As, Sb, Cd, Cr, Se). Further, during the operation, samples can be contaminated with elements from sample containers, or analytes can be lost through sorption. Combustion temperatures are in the range 500–550°C.

Loss of analyte can also occur during *drying* at much lower temperatures (60–90°C), but such losses are generally much smaller.

It seems that the most reliable general method of dehydration is lyophilization.

After dehydration, homogenization of biological samples is often needed because trace elements are unevenly distributed in most materials. Precautions must be taken so that equipment and procedures do not introduce contaminants into the sample. Plastics and high-purity silica instruments are recommended. In the brittle fracture technique, homogenization is done in a Teflon container after freezing the sample with liquid nitrogen, and in this way, soft tissues as well as bones can be handled [12].

Plant materials should be carefully washed free from sands, soils and other physical artifacts. After draining of the wash water, leaves, stems, seeds and fruits must be dried. Undried components are homogenized with a blender with glass walls and high-quality stainless-steel blades. Blending is best done in an inert atmosphere because intensive aeration during homogenization can cause loss of oxidizable components.

Dry biological samples (seeds, grains) are generally ground in mills and sieved through fine-mesh screens to obtain uniformly sieved material. The sieved material can then be pressed into tablets for direct analysis by XRF or activation analysis [13].

Human and animal tissues are generally cut into thin sections or ground in mills. Sectioning, grinding and blending is repeated when necessary. In some cases samples are mineralized by dry or wet ashing techniques. Here again, though dry ashing may be more convenient, it can cause serious loss of volatile materials.

In many cases, simple dry ashing or combustion of biological materials will not completely remove organic matter, and wet ashing with strong oxidizing agents may be required. Alternative procedures in such cases include evaporation to dryness with nitric or perchloric acids, or with mixtures of oxidizing agents (e.g. $H_2O_2+HNO_3+HClO_4$).

In the case of liquid biological materials, mineralization is sometimes required to destroy organic materials which can bind the trace elements in forms that make preconcentration or determination impossible. Mineralization is also used to prolong the holding times of biological fluids that are more unstable (blood, milk, urine, etc.).

3.5 PREPARATION OF SAMPLES FROM SEDIMENTS, SOILS AND SLUDGES

Depending on the analytes present and analytical techniques used, (e.g. metals, and XRF analysis), determinations can sometimes be made without pretreatment. In many other cases the samples must be dissolved or the analytes leach from the solid matrix.

Usually water will dissolve only a small fraction of the solid matrix (chlorides, nitrates, nitrites, some sulphates, some humic materials, fatty acids and sugars and alkali metals not bound in a silicate matrix). Leaching agents for the determination of nutrients (N, P, K, etc.) and some trace elements include dilute acids (HNO_3 or HCl), dilute potassium sulphate solutions and 1% citric acid, acetic acid/acetate buffer, and lactate [14].

In some cases, leaching is not adequate and more complete dissolution of the solid matrix is necessary. Methods used include [14]:

— fusion with sodium carbonate,
— destruction of a silicate matrix with sulphuric and hydrofluoric acids,
— combustion at 550°C,
— wet ashing with strong oxidizing agents (HNO_3, $HClO_4$).

Sludges must sometimes be homogenized with a blender, and it is recommended to use as short a blending time and as slow a blade speed as possible.

REFERENCES

[1] *NBS Standard Reference Materials*, NBS Special Publication 260, 1977 (Annual supplements), Washington, D.C.
[2] *1978–1979 Programme AQCS*, IAEA, Vienna, 1978.
[3] H. J. M. Bowen, *Analyst*, 1967, **92**, 124.
[4] H. J. M. Bowen, *J. Radioanal. Chem.*, 1974, **19**, 215.

[5] H. J. M. Bowen, *At. Energy Rev.*, 1975, **13**, 451.
[6] M. Ördögh, *J. Radioanal. Chem.*, 1978, **46**, 41.
[7] R. Mavrodineanu, J. R. Baldwin and J. K. Taylor, *Final Rep.*, NBS, NBSIR 73-256, Washington, D.C., 1973.
[8] R. Perry and R. J. Young (eds.), *Handbook of Air Pollution Analysis*, Chapman & Hall, London; Wiley, New York, 1977.
[9] P. Schutyser, A. Govaerts, R. Dams and J. Hoste, *J. Radioanal. Chem.*, 1977 **37**, 651.
[10] T. Braun, J. D. Navratil and A. B. Farag, *Polyurethane Foam Sorbents in Separation Science,* CRC Press, Boca Raton, 1985.
[11] J. Minczewski, J. Chwastowska and R. Dybczyński, *Separation and Preconcentration in Inorganic Trace Analysis*, Ellis Horwood, Chichester, 1982.
[12] G. V. Iyengar, *Radiochem. Radioanal. Lett.*, 1976, **24**, 35.
[13] U. S. Saleh, *J. Radioanal. Chem.*, 1982, **74**, 191.
[14] *Standard Methods for the Examination of Water and Wastewater*, 16th Ed., American Public Health Assoc., Washington, D.C., 1986.

4

Analysis of radioactive environmental samples

As indicated in Chapter 1, one of the major concerns of nuclear environmental chemical analysis is the study of radioactivity in the environment — its *nature*, occurrence, behaviour and control. The foundation of such studies is the careful measurement of radioactivity in a variety of environmental samples. Such measurements can be divided into two major types:

(1) measurements on systems of naturally occurring radionuclides where objectives include the determination of identity and amounts of natural radionuclides, and also the use of this information to determine the presence and amounts of the related elements in environmental samples (e.g. U, Th, K).

(2) measurements of man-made (artificial) radionuclides, including determination of identity and amounts (e.g. fission products, fall-out, radioactive spills, leakage from waste disposal sites or other facilities).

4.1 ANALYSIS OF NATURAL RADIOACTIVE MATERIALS

Naturally occurring radionuclides are comprised of two basic groups: primordial and cosmogenic.

Primordial radionuclides are associated with the formation of the earth. A list of these is given in Table 4.1 [1]. The radioactive isotopes of the elements of higher atomic numbers are members of one of four possible decay series:

Thorium series $4n$ $^{232}Th \rightarrow {}^{208}Pb$
Neptunium series $4n + 1$ $^{241}Cm \rightarrow {}^{209}Bi$
Uranium series $4n + 2$ $^{238}U \rightarrow {}^{206}Pb$
Actinium series $4n + 3$ $^{235}U \rightarrow {}^{207}Pb$.

The neptunium series, because of the relatively short half-life of its parent, is no longer present in nature. The decay sequences of the uranium, actinium

Table 4.1 — Primordial radionuclides. (Reproduced from [1] by permission of the copyright holders, Elsevier Science Publishers, Amsterdam).

Nuclide	Half-life (years)	% Isotopic abundance	Emissions, decay†	Energy of primary emission (MeV)	Specific activity (Ci/g element)
^{40}K	1.3×10^{9}	0.0118	β, γ, ε	1.32	8.3×10^{-10}
^{50}V	$\sim6\times10^{14}$	0.24	β, γ, ε	—	$\sim2.8\times10^{-14}$
^{87}Rb	4.7×10^{10}	27.85	β	0.27	2.5×10^{-8}
^{115}In	6×10^{14}	95.72	β	0.6	5.0×10^{-12}
^{138}La	1.1×10^{11}	0.089	β, γ, ε	0.205	2.1×10^{-12}
^{142}Ce	5×10^{15}	11.07	α	1.5	5.7×10^{-14}
^{144}Nd	$\sim5\times10^{15}$	23.85	α	1.8	$\sim1.2\times10^{-13}$
^{147}Sm	1.06×10^{11}	14.97	α	2.24	3.38×10^{-9}
^{148}Sm	1.2×10^{13}	11.24	α	2.14	2.24×10^{-11}
^{149}Sm	$\sim4\times10^{14}$	13.83	α	1.84	$\sim8.2\times10^{-13}$
^{152}Gd	1.1×10^{14}	0.200	α	2.15	4.15×10^{-12}
^{174}Hf	4.3×10^{15}	0.18	α	2.5	8.4×10^{-14}
^{176}Lu	3.6×10^{10}	2.59	β, γ	0.42	1.47×10^{-10}
^{187}Re	7×10^{10}	62.93	β	0.008	1.73×10^{-8}
^{190}Pt	7×10^{11}	0.0127	α	3.11	3.33×10^{-13}
^{192}Pt	$\sim10^{15}$	0.78	α	2.6	$\sim1.4\times10^{-14}$
^{204}Pb	1.4×10^{17}	1.48	α	2.6	1.83×10^{-16}
^{238}U	(series)*				
^{235}U	(series)*				
^{232}Th	(series)*				

†ε = electron capture.
*See Figs. 4.1–4.3.

and thorium series are given in Figs. 4.1–4.3 [1]. Equilibrium quantities of the members of these series are given in Tables 4.2 and 4.3 [2].

Cosmogenic radionuclides are produced continuously as a result of nuclear reactions in the environment, such as interaction with cosmic radiation, spallation and natural fission. A partial list is presented in Table 4.4 [1].

Determinations are usually relative; i.e. samples are compared with standards. For comparative measurements, it may be adequate to make relative measurements without standards, with an instrument calibrated for the purpose.

Direct measurements may be complicated by the phenomenon of self-absorption, especially in the case of low-energy β or α radiation. Two approaches can be used to minimize self-absorption errors:

(1) Measurement of saturation layers, viz. layers which are thicker than the penetration range of the measured radiation, (e.g. >0.1 mm for most α- particles).
(2) Measurement of infinitesimally thin layers; layers which are thin enough for self-absorption to be negligible (layer "density" <0.1 mg/cm^2.

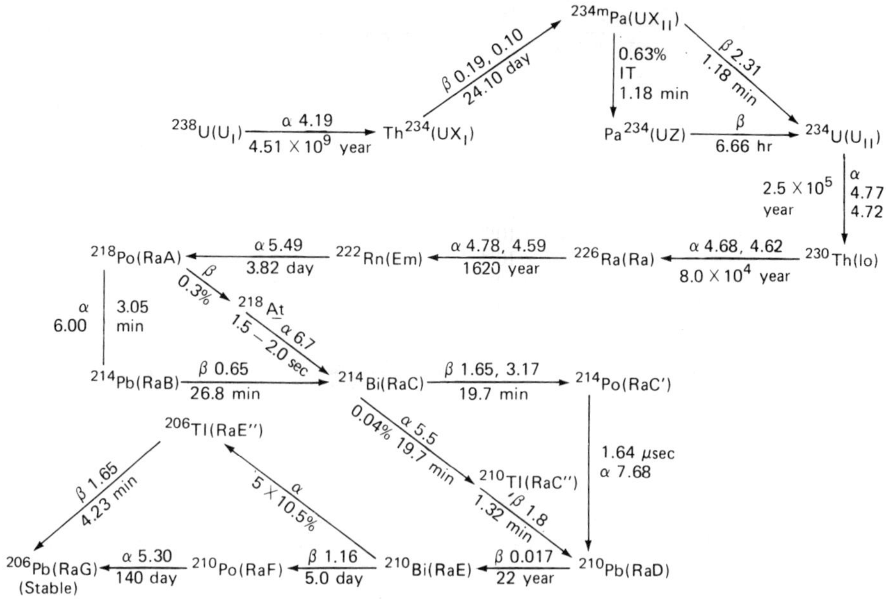

Fig. 4.1 — Uranium series decay scheme (energies in MeV). (Reproduced from [1] by permission of the copyright holders, Elsevier Science Publishers, Amsterdam).

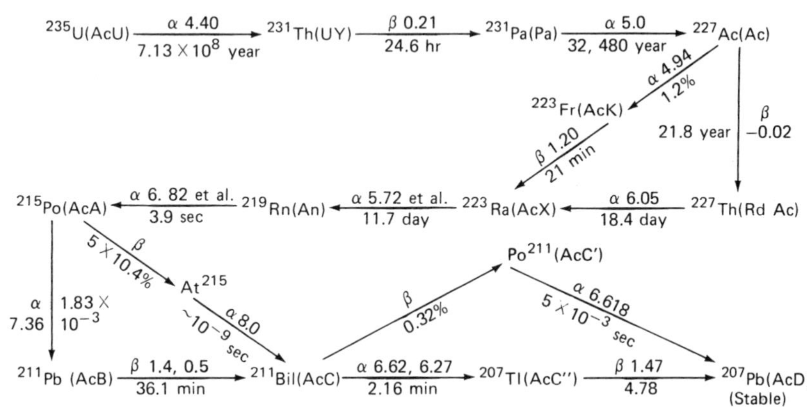

Fig. 4.2 — Actinium series decay scheme (energies in MeV). (Reproduced from [1] by permission of the copyright holders, Elsevier Science Publishers, Amsterdam).

4.1.1 Determination of natural radioactive elements

Naturally occurring radioactive elements can be determined by measuring the activity of appropriate radioisotopes. These measurements generally have remarkable sensitivity and are particularly suitable in cases of short

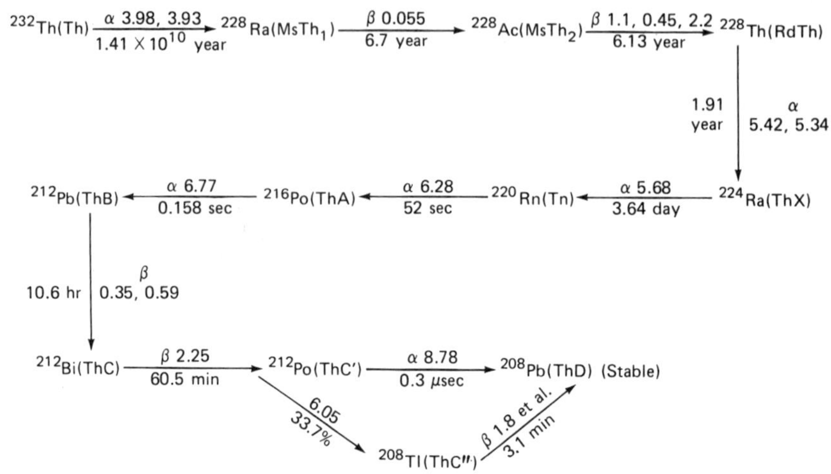

Fig. 4.3 — Thorium series decay scheme (energies in MeV). (Reproduced from [1] by permission of the copyright holders, Elsevier Science Publishers, Amsterdam).

Table 4.2 — Equilibrium quantities of members of the thorium decay series

Nuclide	Half-life (sec)	Amount (g/g Th)
Th	4.38×10^{17}	1.0
MsTh1	2.11×10^{8}	4.81×10^{-10}
MsTh2	2.21×10^{4}	5.05×10^{-14}
RdTh	5.99×10^{7}	1.37×10^{-10}
ThX	3.14×10^{5}	7.17×10^{-13}
Tn	54.5	1.24×10^{-16}
ThA	0.158	3.61×10^{-19}
ThB	3.82×10^{4}	8.72×10^{-14}
At	3×10^{-4}	6.85×10^{-22}
ThC	3.63×10^{3}	8.29×10^{-15}
ThC′ (66.3 %)	3.0×10^{-7}	4.51×10^{-25}
ThC″ (33.7 %)	186	1.43×10^{-16}

half-lives. The use of radioisotopes with long half-lives is less suitable, particularly as regards accuracy and sensitivity; even in this case, however, radiometric methods are usually simpler and more rapid than conventional procedures.

The basic assumption is that a radioelement has at least one isotope which can be measured; hence the isotopic composition must be known. If an element is known to have a constant isotopic composition, measurements and calculations are relatively straightforward (e.g. K, Rb, Sm, Lu, Re, Fr).

When an element has more than one radioisotope, determinations and

Table 4.3 — Equilibrium quantities of members of the uranium and actinium decay series

Nuclide	Half-life (sec)	Quantity (g/g Ra)	Quantity (g/g U)
UI	1.42×10^{17}	2.83×10^{6}	1
UX1	2.12×10^{6}	4.23×10^{-5}	1.49×10^{-11}
UX2 (99.85%)	68.4	1.36×10^{-9}	4.82×10^{-16}
UZ (0.15%)	2.41×10^{4}	7.20×10^{-10}	2.50×10^{-16}
UII	7.35×10^{12}	1.47×10^{12}	5.18×10^{-5}
Io	2.62×10^{12}	52.3	1.85×10^{-5}
Ra	5.01×10^{10}	1	3.53×10^{-7}
Rn	3.30×10^{5}	6.59×10^{-6}	2.32×10^{-12}
RaA	183	3.65×10^{-9}	1.29×10^{-15}
RaB	1.61×10^{3}	3.21×10^{-8}	1.13×10^{-14}
RaC	1.18×10^{3}	2.36×10^{-8}	8.31×10^{-15}
RaC′ (99.96%)	1.60×10^{-4}	3.0×10^{-15}	1.06×10^{-21}
RaC″ (0.04%)	79.2	6.30×10^{-12}	2.28×10^{-19}
RaD	6.94×10^{8}	1.38×10^{-2}	4.89×10^{-9}
RaE	4.32×10^{5}	8.62×10^{-6}	3.04×10^{-12}
Po (RaF)	1.21×10^{7}	2.41×10^{-4}	8.52×10^{-11}
AcU (0.71%)	2.24×10^{16}	2×10^{4}	7.10×10^{-3}
Pa	1.01×10^{12}	0.88	3.12×10^{-7}
Ac	6.62×10^{8}	5.8×10^{-4}	2.06×10^{-10}
UY	9.20×10^{4}	8.2×10^{-8}	2.9×10^{-14}
RaAc	1.63×10^{6}	1.44×10^{-6}	5.1×10^{-12}
AcX	9.7×10^{5}	8.5×10^{-7}	3.0×10^{-13}
An	3.92	3.46×10^{-12}	1.22×10^{-18}
AcA	1.83×10^{-3}	1.6×10^{-15}	5.72×10^{-22}
AcB	2.17×10^{3}	1.88×10^{-9}	6.76×10^{-16}
AcC	130	1.16×10^{-10}	4.08×10^{-17}
AcC′ (0.32%)	0.52	1.49×10^{-15}	5.27×10^{-22}
AcC″ (99.68%)	286	2.52×10^{-10}	8.95×10^{-17}

Table 4.4 — Major cosmic-ray-activated radionuclides. (Reproduced from [1] by permission of the copyright holders, Elsevier Science Publishers, Amsterdam).

Nuclide	Half-life	Emission, decay†	Energy of primary emission (MeV)	Production rate (atom. cm^{-2}. sec^{-1})
^{3}H	12.26 years	β	0.0181	0.12—1.3
^{7}Be	53 days	ε, γ	—	0.021—0.035
^{10}Be	2.7×10^{6} years	β	0.56	0.04—0.1
^{14}C	5730 years	β	0.156	2—2.6
^{22}Na	2.58 years	β^{+}, ε, γ	0.54	—
^{32}SI	~650 years	β	0.1 (to ^{32}P)	2×10^{-4}
^{32}P	14.3 days	β	1.71	1×10^{-4}
^{33}P	25 days	β	0.25	1×10^{-4}
^{35}S	86.7 days	β	0.168	2×10^{-4}
^{36}Cl	3×10^{5} years	β, ε	0.71	—
^{39}Cl	55 min	β, γ	1.91, 2.18, 3.43	—

†ε = electron capture.

data analysis are generally more complex because the isotopes may differ considerably in half-life, especially if a series is involved, e.g. Ra, Th, Po, Rn, Ac, Pa and U. One possibility is to make measurements after the decay of the short-lived radionuclides, but this may require long waiting times. In favourable cases, it is more convenient to measure the activity of decay products (e.g. radon, thoron, actinon), or correct the measurements of the short-lived radioisotopes after determination of the isotopic composition [3].

4.1.2 Emanometric analysis

Emanometric methods are radioanalytical methods which use measurement of radioactive isotopes of inert gases for the determination of appropriate elements [4]. In general, these methods have good sensitivity and accuracy, but also disadvantages; they require the handling of radioactive gases, the analysis is usually slow (up to several days) and reliable results require the establishment of radioactive equilibrium in the detection unit.

A good example is the use of the radon isotopes ^{226}Rn, ^{222}Rn and ^{219}Rn to determine radon, thorium, radium and actinium. Indirect determinations (e.g. use of Rn to determine Ra) require both components to be in equilibrium, or present in a known ratio.

In general, these methods require the removal of the inert gas from the sample, and transfer to a suitable detector. Solid samples must be converted into liquid form such as a solution, or a melt by fusion with sodium carbonate or borax. Once the sample is in liquid form, the inert gas may be swept out by bubbling air through the solution into an evacuated ionization chamber; passing a continuous stream of air through the sample and detector (useful for short-lived radioisotopes); spraying the liquid by means of a small air-screw or shaking the liquid sample with air.

4.1.2.1 Determination of individual elements

Radon and radium. In nature, radon is most often associated with uranium deposits, since some of its isotopes are formed as part of the natural decay series, see Figs. 4.1–4.3. In isolated samples, radon will reach radioequilibrium in about 3 hr, and the total rate of α-particle emission will be 3 times that for radon itself. This is because for each α-particle from radon decay, approximately one from ^{218}Po and one from ^{214}Po will also be emitted. Since the half-life of ^{210}Po (22 yr) is long compared to analysis times, its decay can be neglected. However, since analysis times are still significantly long, the decay of the radon itself must be taken into account.

The determination of radium by the emanation method is based on the production of radon as a decay product of radium:

$$^{226}\text{Ra} \xrightarrow[1620\ \text{yr}]{\alpha} {}^{222}\text{Rn}$$

$$^{223}Ra \xrightarrow[11.7\ d]{\alpha} {}^{219}Rn$$

$$^{224}Ra \xrightarrow[3.6\ d]{\alpha} {}^{220}Rn$$

From a liquid sample, the radon is first flushed out with a carrier gas. The end of the flushing is taken as the starting time for fresh radon accumulation in the sample. After a suitable time, the radon reaches equilibrium in the sample and is then transferred to the counting device by a second flushing with a carrier gas (e.g. 15% H_2 + 85% N_2). After the radon isotopes have reached radioequilibrium with their daughters in the detector (3–4 hr), a measurement is made. The measured radon activity is used to calculate the amount of radium in the sample. The method can detect as little as 10^{-12} g of radium.

An example of a radon bubbler is shown in Fig. 4.4 [5]. Since radon

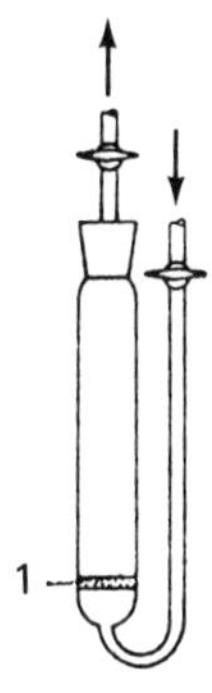

Fig. 4.4 — Radon bubbler, 1, fritted disk.

isotopes are gaseous and α-particle emitters, direct introduction into a specially designed ionization chamber is an effective method of measurement. A diagram of such a device is shown in Fig. 4.5 [2]. Alternatively, an α-scintillation cell coupled to a photomultiplier can be used. A typical cell is shown in Fig. 4.6 [5, 6].

Actinium. The nuclide ^{219}Rn can be used for the determination of actinium:

$$^{227}Ac \xrightarrow{\alpha} {}^{223}Ra \xrightarrow{\alpha} {}^{219}Rn$$

The ^{219}Rn activity of a sample is compared to that of a standard with a

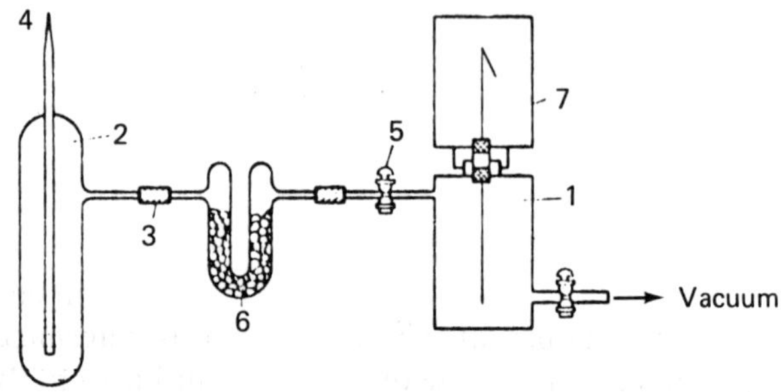

Fig. 4.5 — Emanation equipment for the determination of radium: 1, ionization chamber; 2, ampoule for radon growth; 3, 4, tubes; 5, stopcock; 6, $CaCl_2$; 7, ionization chamber system.

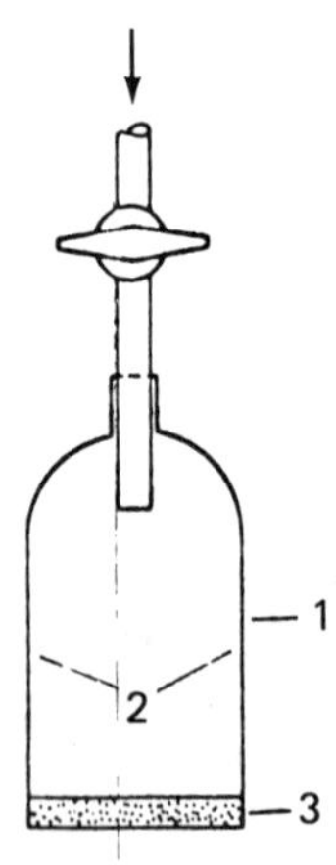

Fig. 4.6 — Alpha-scintillation cell: 1, forging metal; 2, phosphor layer; 3, glass window coated with SnO_2.

known content of ^{223}Ra, which is prepared from a uranium ore containing a negligible amount of thorium. In such cases, it can be assumed that the ratio of $^{235}U/^{238}U$ is 0.007. Since the decay products are short-lived, equilibrium is attained within a few minutes. Amounts of actinium as small as 10^{-14} g have been determined by this method [7].

Thorium. The only gaseous product in the thorium decay series is thoron, which has a very short half-life (56 sec). It can be used to determine ^{220}Rn, and also its parent, ^{232}Th. Thorium is determined by comparison with a standard, but because of the short half-life of ^{220}Rn, decay corrections must be made. Naturally occurring thorium cannot be used as a standard because of the presence of ^{230}Th. If the sample is not in equilibrium, corrections must be made for the presence of long-lived isotopes such as ^{228}Ra and ^{228}Th.

The ^{220}Rn activity must be measured dynamically in the stream of carrier gas. A device for such measurements is shown in Fig. 4.7 [8]. The sample is

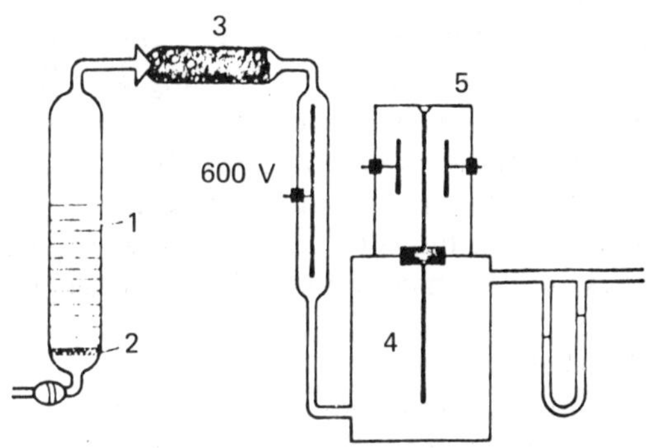

Fig. 4.7 — Emanation equipment for the determination of thoron: 1, purge system; 2, fritted glass disk; 3, filter with glass wool; 4, ionization chamber and electrometer.

placed in a container (1), air dispersed by the fritted glass disk (2) is used as a purge, and the air + thoron mixture is filtered (3) before entering the ionization chamber (4). The flow of carrier gas should be relatively high so that the time of passage from the purge system to the ionization detector is short compared to the average life of the thoron, and must be kept constant during measurements.

A solution of known thorium content (1–10 mg) is used as a standard.

Measurement of ^{222}Rn and ^{220}Rn in mixtures. The determination of ^{222}Rn and ^{220}Rn in mixtures is relatively simple because of the difference in half-life values (3.8 day and 56 sec). One technique consists of measuring the combined activity of the nuclides as a function of time. The total activity at first decreases, by decay of ^{220}Rn. After reaching a minimum value, it rises moderately with time because of the growth of daughter products of ^{222}Rn.

The ratio of the active components may be determined from an activity *vs.* time curve, as shown in Fig. 4.8 [9]. The ratio OT:OR is used.

4.1.3 Analysis of environmental samples

4.1.3.1 Air

The natural radioactivity in air is due mainly to radioactive emanations (radon, thoron, actinon) and by their deposits which produce radioactive aerosols. In the continental atmosphere, the normal concentration of radon is of the order of 10^{-6} Bq/cm^3, and for thoron, 10^{-8} Bq/cm^3. Air over the oceans generally has concentrations two orders of magnitude lower. The activity of air in soils is about three orders of magnitude higher than that in the open atmosphere (10^{-3} Bq/cm^3). Air in mines, particularly those with uranium deposits, will have higher concentrations.

For the analysis of air samples, a number of approaches are possible.

(1) Direct emanometric determination with large-volume (20-l.) differential ionization chambers which permit measurements at activity levels around 10^{-7} Bq/cm^3.
(2) Preconcentration of radioactive emanation by condensation, adsorption on activated charcoal, cooling with liquid air, or by sorption in cooled organic solvents (e.g. toluene, carbon disulphide). After collection, the radioactive gases can be purged from the collectors and transferred to an appropriate ionization chamber.
(3) Capture (e.g. filtration) of aerosols and measurement of their activity with scintillation counting or by autoradiography.

4.1.3.2 Water

The amount of natural radioactive materials in water varies considerably, depending on the nature, location and history of the water. A list of the major natural and artificial radionuclides and their origin is given in Table 4.5 [10]. The principal natural radionuclides are ^{238}U, ^{232}Th, ^{235}U and their decay products, such as ^{226}Ra and ^{222}Rn from ^{238}U, ^{228}Ra and ^{224}Ra from ^{232}Th [11].

Natural waters may contain of the order of 1.5 kBq/l. ^{222}Rn; 0.2 Bq/l. ^{226}Ra; 30 μg/l. U and Th. Concentrations can be substantially higher in water draining from uranium mines, e.g. 0.2 MBq/l. ^{222}Rn; 0.3 kBq/l. ^{226}Ra; 90 mg/l. U.

If we discount the presence of U and Th, the contribution of which to the total activity is usually relatively small, we can differentiate between *temporary* radioactivity caused by Rn, and *permanent* activity caused by Ra.

For the determination of Rn, the emanation can be flushed to an ionization chamber and measured as described above.

Radium can be determined by one of two methods [12]. First, *α-counting of a barium–radium sulphate precipitate* that has been purified. The method will include all α-emitting radium isotopes present and is designed primarily for clear water. Sewage and industrial wastes can also be studied, but steps must be taken to remove organic matter and other interferences. Ashing should be avoided or a fusion may become necessary. Lead and barium

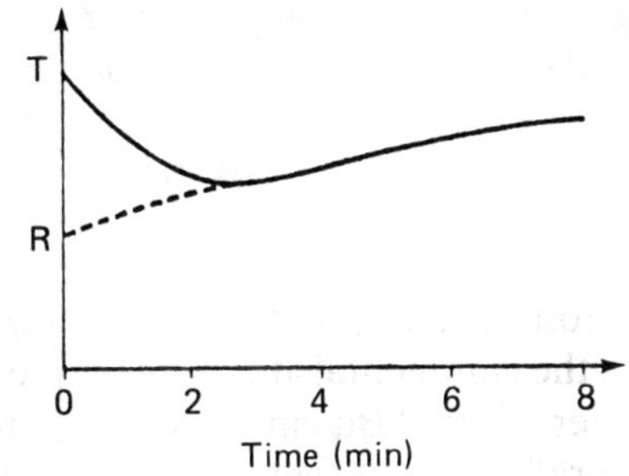

Fig. 4.8 — Activity *vs*. time for a mixture of ^{222}Rn and ^{220}Rn.

Table 4.5 — Radionuclides in water

Radionuclide	Half-life	Nuclear reaction, description, source
Naturally occurring and from cosmic reactions		
Carbon-14	5730 years	$^{14}N(n, p)\,^{14}C$,† thermal neutrons from cosmic or nuclear-weapon sources, reacting with N_2
Silicon-32	~650 years	$^{40}Ar(p,x)\,^{32}Si$, nuclear spallation (splitting of the nucleus) of atmospheric argon by cosmic-ray protons
Potassium-40	$\sim 1.3 \times 10^9$ years	0.0118% of natural potassium
Naturally occurring from ^{238}U series		
Radium-226	1620 years	Diffusion from sediments, atmosphere
Lead-210	21 years	$^{226}Ra \rightarrow$ 6 steps $\rightarrow {}^{210}Pb$
Thorium-230	75,200 years	$^{238}U \rightarrow$ 3 steps $\rightarrow {}^{230}Th$ produced *in situ*
Thorium-234	24 days	$^{238}U \rightarrow {}^{234}Th$ produced *in situ*
From reactor and weapons fission		
Strontium-90	28 years	These are the fission-product radioisotopes of greatest significance because of their high yields and biological activity
Iodine-131	8 days	
Caesium-137	30 years	
Barium-140	13 days	The isotopes from barium-140 to krypton-85 are listed in generally decreasing order of fission yield
Zirconium-95	65 days	
Cerium-141	33 days	
Strontium-89	51 days	
Ruthenium-103	40 days	
Krypton-85	10.3 years	
Cobalt-60	5.25 years	From non-fission neutron reactions in reactors
Manganese-54	310 days	From non-fission neutron reactions in reactors
Iron-55	2.7 years	$^{56}Fe(n, 2n)\,^{55}Fe$, from high-energy neutrons acting on iron in weapon hardware
Plutonium-239	24,300 years	$^{238}U(n, \gamma)\,^{239}Pu$, neutron capture by uranium

†This notation denotes the isotope nitrogen-14 reacting with a neutron, n, giving off a proton, p, and forming the isotope carbon-14; other nuclear reactions may be similarly deduced from the notation shown. (Note that x represents nuclear fragments from the spallation reaction.)

carriers are added to the sample, containing an alkali-metal citrate, then sulphuric acid is added to precipitate radium, barium and lead as sulphates. The precipitate is purified by washing with nitric acid, redissolving in alkaline EDTA and reprecipitating as radium–barium sulphate by pH adjustment to 4.5. The precipitate is then α-counted to determine radium. Proportional and α-scintillation counters can be used. A variation of the method consists in sorption of radium onto crystals of barium sulphate dispersed in a gelatine emulsion, and counting. Secondly, *measurement of* ^{222}*Rn produced from* ^{226}*Ra*. Radium in water is concentrated and separated by co-precipitation with an excess of barium, as the sulphate. The precipitate is treated to remove silicate and decompose insoluble radium compounds, heated with phosphoric acid to remove sulphite, and dissolved in hydrochloric acid. The completely dissolved radium is placed in a bubbler and stored for up to 4 weeks for ingrowth of radon. The bubbler is connected to an evacuated system and the radon gas is removed from the liquid by aeration, dried with a desiccant, and collected in a counting chamber. The chamber is usually a dome-shaped scintillation cell coated inside with a silver-activated zinc sulphide phosphor (see Fig. 4.6).

4.2 ANALYSIS OF MAN-MADE RADIOACTIVE MATERIALS

The analysis of natural radioactive materials is relatively straightforward, since it usually involves the determination of a small number of well-known radioisotopes of a few elements. The determination of man-made radionuclides in environmental samples is more complicated, because several radionuclides of a number of elements may be involved, both qualitative and quantitative studies may be needed, and carriers may have to be identified.

4.2.1 Qualitative analysis

The final aim of the qualitative analysis is identification of all radioisotopes present in a sample (identification of all activities and the nuclides responsible). The procedure has two steps.

(1) The characterization of all activities present; e.g. nature and energy of radiation, relative intensities, half-life values.
(2) Identification of the radionuclides responsible for the various activities.

The nature and energy of the radiation can be determined [13] by:

(1) gamma-spectrometry with NaI(Tl) crystals or solid state detectors such as Ge/Li or Si/Li systems,
(2) alpha-spectrometry with surface barrier detectors,
(3) beta-spectrometry with silicon semiconductor detectors (surface barrier or Si/Li type),
(4) absorption measurements.

Half-life values can be determined from decay curves showing activity *vs*. time, and these values must be correlated with the types of radiation found. Once this correlation has been made, appropriate tables [14] can be used for nuclide identification. If an observed activity cannot be identified, or if it is associated with a "new" nuclide, it is sometimes labelled only by half-life; e.g. 18.3 day activity, or 5.3 sec activity.

4.2.2 Quantitative analysis

The aim is to measure an amount of radioactivity in either a relative or absolute sense. Common units are Becquerels, Becquerels per unit mass or Becquerels per unit volume. In special cases, the activity is expressed as a mass of the radioactive isotope.

4.2.2.1 Absolute determination of activity

The following equation is generally used to express the relationship between absolute activity, which is a disintegration rate, and a detector signal:

$$A = I/\xi, \text{ where} \tag{4.1}$$

A = the absolute activity in Becquerels (disintegrations per second),
I = the signal from the detector, e.g. counts per unit time, current or area under a spectral peak,
ξ = overall efficiency of the sample–detector system.

The overall efficiency factor is usually thought of as being a product of terms such as:

$$\xi = \omega K S q \xi', \text{ where} \tag{4.2}$$

ω = a geometric factor, usually the solid angle subtended by the detector, and expressed as a fraction of 4π steradians,
K = absorption losses in air and the window of the detector,
S = self-absorption in the sample and sample cover,
q = back-scattering from the sample support,
ξ' = internal efficiency of the detector/electronic system.

The absolute determination of activity is possible only in those cases where all the factors in the equations above are known or can be determined, which is usually quite difficult. For α and β emitters, it is easier to use experimental conditions such that some of the factors can be assumed to be nearly unity, for example uniformly thin samples ($S = 1$) on an infinitely thin support ($q = 1$), inside a relatively large detector ($K\xi'\omega = 1$). An example is the use of a 4π internal proportional counter.

Another option is simply to evaluate ξ for a given experimental arrangement by use of an appropriate standard of known activity. Strictly speaking, ξ evaluated in this way is valid only for the type and energy of radiation emitted by the standard.

4.2.2.2 *Relative determination of activity*

Relative measurements of activity are comparisons of the activity of a sample with that of some standard or reference sample under the same experimental conditions. Units will usually be counts per minute, area under a curve, or a number of scale divisions.

4.2.3 Differentiation of radioactive nuclides in mixtures

The analysis of mixtures of radionuclides may present two different tasks.

(1) The detection and determination of a single radionuclide with all others considered as interferences.
(2) The detection and determination of several or all radioactive components.

If these tasks are managed without chemical or physical separations, we generally use the term *resolution* (differentiation). As indicated earlier, resolution is attempted on the basis of half-lives and the nature and energy of the radiation. Spectral analysis, growth/decay curve analysis and absorption studies are all used. High-resolution solid-state detectors, multichannel pulse-height analysers and computerized data analysis can be very helpful. Computer codes exist for the resolution (deconvolution) of complex spectra [15,16].

In favourable cases, if the half-lives of interferences are relatively short, they can be left until they have decayed, and the simpler spectrum left can then be resolved.

If resolution is not possible because of spectral complexity or some other reason, then chemical/physical separations are necessary. Nuclear data are collected for the separated fractions, and resolution is then generally simpler.

4.2.3.1 *Analysis of fission products*

Fission products are nuclides produced by primary and secondary processes associated with fission, generally of the uranium isotopes ^{235}U and ^{238}U. The mixture of products contains, besides the final stable nuclei, more than 200 radionuclei with half-lives ranging from fractions of seconds to many years. Their atomic numbers range from 30 to 65 (Zn–Tb) and mass numbers from 72 to 161.

The probability of producing a given nuclide in a fission event is called the *fission yield* (per cent). The sum of yields for all products is 200% because each event produces two fragments. An example of a fission yield curve is given in Fig. 4.9 [17].

The main task in the analysis of a mixture of fission products is the identification of nuclides and the determination of yields. A combination of instrumental and chemical technique is required and the work is laborious and difficult.

Determination of total radioactive strontium and ^{90}Sr. Strontium-90 is one of the most hazardous of the fission products, because of its long half-life (28

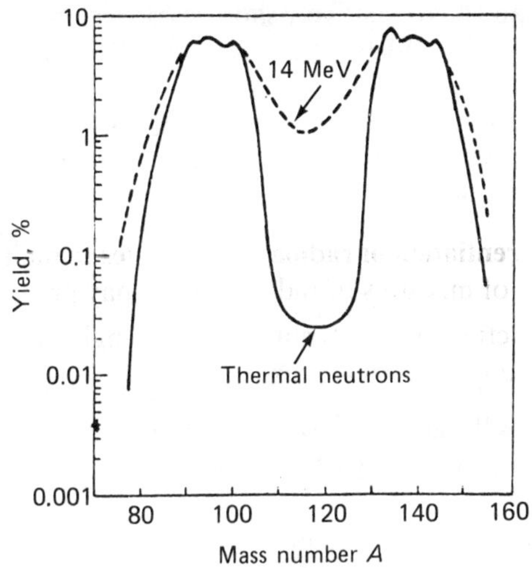

Fig. 4.9 — ^{235}U fission yield for thermal-neutron induced fission. The dashed line indicates the yield when fission is induced by 14-MeV neutrons.

yr), and because strontium is concentrated in bone. For the determination of total radioactive strontium and ^{90}Sr in water, the following procedure is used [12].

A known amount of strontium [as the nitrate, $Sr(NO_3)_2$], is added as carrier to a water sample. The carrier, alkaline-earth and rare-earth metals are precipitated as their carbonates to concentrate the radiostrontium. The strontium is then separated from other radioactive elements and inactive sample solids by precipitation as $Sr(NO_3)_2$ from fuming nitric acid solution, and finally precipitated as strontium carbonate, $SrCO_3$, dried, weighed to determine the carrier recovery, and measured for radioactivity. A correction is applied to compensate for loss of carrier and activity during the purification procedure. A delay in counting will give an increased counting rate due to the ingrowth of ^{90}Y. The total activity will be due to ^{89}Sr and ^{90}Sr.

The amount of ^{90}Sr can be determined by separating and measuring its daughter, ^{90}Y, after equilibrium has been reached. Two procedures can be used. In the first, ^{90}Y is separated by extraction into tributyl phosphate from concentrated nitric acid solution. It is back-extracted into dilute nitric acid and this solution is evaporated to dryness for β-counting.

In the second method, yttrium carrier is added, separated by precipitation as yttrium hydroxide, reprecipitated as yttrium oxalate, and counted.

Counting instruments used include internal gas-flow proportional counters and thin-window proportional or GM counters (preferably with a low-background anti-coincidence system). The counting systems can be calibrated with ^{137}Cs standards.

Radioactive caesium. Upon ingestion, caesium distributes itself throughout soft tissues, and has a relatively short residence time in the human body. The half-lives of ^{134}Cs and ^{137}Cs are 2.0 and 30 yr respectively, and both are β/γ-emitters.

If the activity of caesium is high, it can be determined by direct γ-spectrometry on a large sample (4 l.), or the sample can be evaporated and counted. For lower activity levels, caesium carrier is added and caesium is collected by precipitation as the phosphomolybdate. The precipitate is purified and the caesium is finally precipitated as chloroplatinate, Cs_2PtCl_6, for counting. Measurements are made either by β-counting or by γ-spectrometry [12,18].

Radioactive iodine. Fission products may contain the iodine isotopes from ^{129}I to ^{135}I. Of these, ^{131}I is considered the most significant hazard in drinking water, and the USEPA has set a maximum contaminant level of 3 pCi/l.

Three methods are available for the determination of radioactive iodine in water samples: precipitation, sorption on an anion-exchange resin, or distillation. The precipitation method is preferred because it is simple and requires the least time.

In the precipitation method, iodate carrier is added to the sample and reduced to iodide with sodium sulphite. The iodide is precipitated as silver iodide. The precipitate is dissolved, and purified with zinc powder and sulphuric acid. The iodide is finally precipitated as palladium iodide, PdI_2, for counting in a low-background β-counter, or a β/γ coincidence system [19, 20].

4.2.3.2 Analysis of transuranium nuclides

In the determination of transuranium elements or nuclides, the most important step is separation from the sample matrix. Differences in redox properties are used for the separation of the first four elements in the series (Np,Pu,Am,Cm). Since the higher members exist primarily in the same oxidation state, (III), separation by ion-exchange chromatography is commonly used.

The lighter transuranic elements can be determined by common chemical methods, and trace amounts are usually determined by radiometric methods such as α-spectrometry. The interested reader is referred to standard references for detailed information on specific methods [21–23].

4.2.3.3 Analysis of transplutonium elements

Ion-exchange chromatography is generally used for the separation of the transplutonium elements (Am,Cm,Bk,Cf,Es,Fm,Md,No,Lr). Determinations are usually made directly by α-spectrometry with solid-state detectors. Some elements (Am,Cm,Bk,Cf) also have long-lived isotopes and can be determined by chemical methods such as spectrophotometry. Detailed information can be found in various sources [21–24].

REFERENCES

[1] A. W. Klement, Jr., in *Radioactive Fallout, Soils, Plants, Foods, Man*, E. B. Fowler (ed.), p. 113. Elsevier, Amsterdam, 1965.
[2] S. J. Bresler, *Radioactive Elements* (in Czech), NČSAV, Praha, 1959.
[3] J. Tölgyessy and Š. Varga, *Nuclear Analytical Chemistry*, Vol. II, University Park Press, Baltimore; Veda, Bratislava, 1972.
[4] V. Balek and J. Tölgyessy, *Emanation Thermal Analysis and other Radiometric Emanation Methods*, Akadémiai Kiadó, Budapest, 1984.
[5] *Methods of Radiochemical Analysis*, WHO, Geneva, 1966.
[6] H. F. Lucas, *Rev. Sci. Instrum.*, 1957, **28**, 680.
[7] V. I. Baranov, *Akad. V.I. Vernadskomu k Pyatidessyatiletiyu Nauch. Deyatelnosti*, 1936, **1**, 499; *Chem. Zentr.*, 1938, **I**, 1533.
[8] J. Tölgyessy, *Nuclear Radiation in Chemical Analysis* (in Slovak), SVTL, Bratislava; SNTL, Prague, 1962.
[9] V. Majer, *Radiometric Analysis* (in Czech), Čs. Spol. Chem., Prague, 1949.
[10] S. E. Manahan, *Environmental Chemistry*, 4th Ed., Willard Grant Press, Boston, 1984.
[11] J. D. Hem, *Geol. Survey Water Supply Paper* 1473, U.S. Government Printing Office, Washington, D.C., 1970.
[12] *Standard Methods for the Examination of Water and Wastewater*, 16th Ed., American Public Health Association, Washington, D.C., 1986.
[13] N. Tsoulfanidies, *Measurement and Detection of Radiation*, McGraw-Hill, New York, 1983.
[14] R. L. Heath, *Table of Isotopes, CRC Handbook of Chemistry and Physics*, 66th Ed., CRC Press, Boca Raton, Florida, 1985–86.
[15] *Forist Spectra Unfolding Code*, RSIC Computer Code Collection, PSR-92, ORNL, 1975.
[16] *COOLC and FEDOR Spectra Unfolding Codes*, RSIC Computer Code Collection, RSR-17, ORNL.
[17] G. J. Dilorio, *Direct Physical Mass Yields in Thermal Fission of ^{235}U*, Ph.D. Thesis, University of Illinois at Urbana-Champaign, 1976.
[18] EPA, *Interim Radiochemical Methodology for Drinking Water*, Environmental Monitoring and Support Laboratory, Cincinnati, Ohio, 1976 and updates.
[19] F. P. Brauer, J. J. Kaye, and R. E. Connally, *X-Ray and β-γ Coincidence Spectrometry Applied to Radiochemical Analysis of Environmental Samples*, in Advances in Chemistry Series, No. 93, *Radionuclides in the Environment*, pp. 251–253, American Chemical Society, Washington, D.C. 1970.
[20] F. B. Johns (ed.), *Handbook of Radiochemical Methods*, EPA-680/4-75-001, National Environmental Research Center, Office of Research and Development, USEPA, Las Vegas, 1975 and updates.
[21] Environmental Instrumentation Group, *Instrumentation for Environmental Monitoring*, LBL-1, University of California, Lawrence Berkeley Laboratory, 1972 and updates.
[22] *National Council on Radiation Protection and Measurement (NCRP), Report* No. 62, Washington, D.C., 1979 and updates.
[23] H. L. Krieger and S. Gold, *Procedures for Radiochemical Analysis of Nuclear Reactor Aqueous Solutions*, EPA-R4-73-014, U.S. Environmental Protection Agency, Washington, D.C., 1973 and updates.
[24] B. F. Myasoedov, *Talanta*, 1987, **34**, 31.

5

Isotope Dilution Analysis

The principles of isotope dilution analysis (IDA) were first published 55 years ago by Hevesy and Hobbie [1] and Starik [2]. A significant contribution was made by Růžička and Starý [3–5] when they introduced the substoichiometric modification, which gave greatly increased sensitivity and versatility to the basic method. The theory of IDA, its many variations, and a broad discussion of its application in many fields are provided in the monograph by Tölgyessy *et al.* [6].

The basic principle of IDA is the conservation of activity on dilution, analogous to mass conservation in classical chemical dilution problems. If a radioactive material is diluted with its non-radioactive counterpart, the specific activity (activity per unit mass or volume) of the resulting dilution mixture is related to the original specific activity and the amount of material by a simple conservation equation. This equation can be rearranged in various ways, depending on the conditions of the experiments and solved for the unknown mass of analyte in the sample.

The basic method and its many variants have been used for the determination of trace elements (especially metals) in a wide variety of environmental samples, including aerosols, water, soils and biological materials.

The method can be useful in the following general situation:

- a substance is to be determined in a mixture of similar materials, but a quantitative isolation is impossible;
- the analyte occurs at a low concentration, so that losses such as sorption onto vessel surfaces during the separation procedures are inevitable;
- the analysis must be performed quickly, e.g. because of decay or shifts in equilibria;
- the analyte is part of a large system, and only a portion of it is available, e.g. water in a large living animal.

5.1 VARIATIONS OF IDA

A large number of variations of IDA exist, and there has been some confusion as to proper terminology. Tölgyessy *et al.*, [6] have indicated four different criteria for classification of the variations.

(1) Manner of introducing radioactivity into the system.
(2) Method of determining the quantity of isolated fraction(s).
(3) Number of dilutions.
(4) Relative weights of analyte and diluent.

The four criteria and their resulting classification schemes are given in Fig. 5.1. The manner in which radioactivity is introduced into the system leads to the following types of IDA (the labels refer to Fig. 5.1).

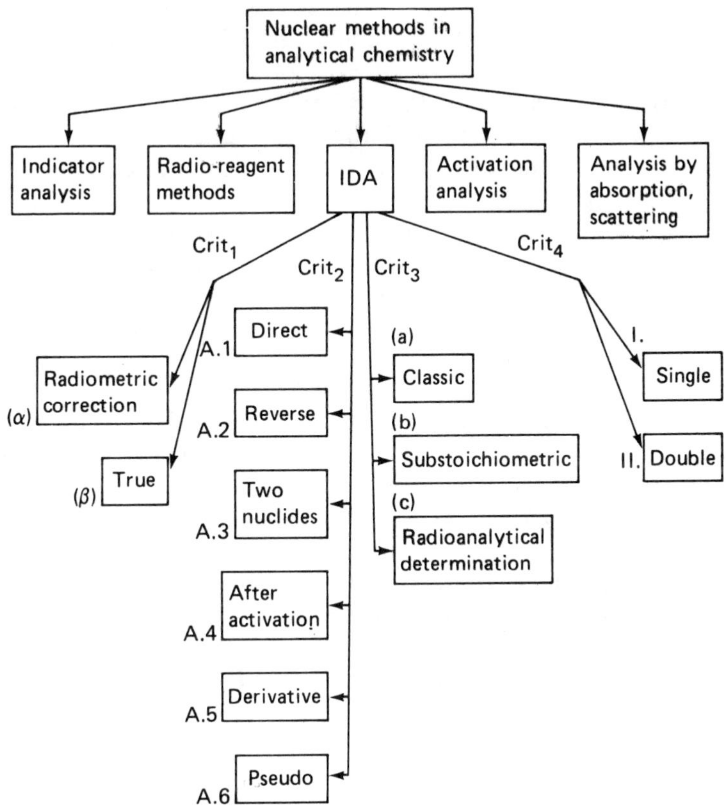

Fig. 5.1 — Classification of nuclear methods in analytical chemistry and IDA. (Adapted from [6] by permission of the copyright holders, Pergamon Press, Oxford).

(A.1) Direct IDA; dilution of the tagged analyte with its non-radioactive analogue.

(A.2) Reverse IDA; dilution of a radioactive sample with stable isotopes.

(A.3) The yield of a separation procedure for a given radionuclide is measured by dilution with a second (isotopic) radionuclide. The activities of the two nuclides must be measured separately.
(A.4) IDA after activation begins with a non-radioactive sample, and the radioactivity of the analyte is induced by a suitable activation technique. The analyte is then diluted with its non-radioactive counterpart.
(A.5) Derivative dilution analysis. The analyte is originally non-radioactive, but is made so by reaction with a radioactive reagent.
(A.6) Pseudoisotopic dilution analysis. The mixed substances are not isotopic (different elements), but are chemically similar.

If the variations are classified according to the means used to determine the amount of isolated fraction, three variations can be distinguished.

(a) Classic IDA, where the amount of isolated fraction is determined by classical analytical methods.
(b) Substoichiometric IDA, where equal but usually unknown amounts of material are separated from two solutions.
(c) The amount of isolated material is determined by an auxiliary method.

IDA can also be subdivided according to the number of dilution steps in the procedure. (I) In single IDA, one dilution step is used, and (II) in double IDA, at least two are necessary.

Comparison of the amounts of unknown analyte and diluting material leads to the variations:

(α) IDA with radiometric correction for yield or isolation. The mass of the radioactive diluent can be neglected compared to the mass of analyte in the sample.
(β) True IDA, where the mass of analyte and tag are similar.

The reader is referred to the monograph by Tölgyessy *et al.* [6] for detailed discussion of all these variants. Since we are dealing with applications in environmental chemical analysis, we will discuss in broad terms only classical IDA, substoichiometric IDA and the more recent development, sub-superequivalence IDA.

5.2 CLASSICAL IDA

The classical technique uses the comparison of the specific activity of a radioactive tracer before and after mixing with a non-radioactive compound which is to be determined (the analyte). In other words, the radioactive tracer is 'diluted' with its non-radioactive counterpart in the sample. The dilution causes a change in the specific activity of the added tracer, which can be measured and used to calculate the amount or concentration of the component of interest in the sample. The basic concept is described in Fig. 5.2, which includes the related mathematical equations.

It should be noted that the condition $E_s = E_x$ refers to equal counting efficiency of samples and standards, and is crucial for reliable results. The term R_x/M_x should be independent of mass, and therefore can be measured on any pure portion of the compound separated from the sample, assuming that the sample is homogeneous, and that sufficient material is isolated for the reliable measurement of R_x and M_x. It should be noted that with this technique, it is not necessary to separate the analyte quantitatively from the sample, provided that the final counting sample is pure.

A number of variations of classical IDA have been developed, including:

— direct IDA, or single IDA, where the non-radioactive sample is diluted with a radioactive tracer,
— reverse IDA, where a radioactive substance is diluted with a stable one,
— derivative IDA, where the analyte is originally non-radioactive, but is made radioactive by a stoichiometric reaction with a radioactive reagent,
— double isotope dilution, where two radioactive isotopes of the same element are used,
— IDA after activation, where the radioactivity of the analyte is induced by an appropriate activation technique,
— pseudo IDA, in which the analyte is not the same element as the tracer, but has adequately similar chemical properties.

The advantages of classical IDA include:

— wide range of applicability; the method has been used for several elements and many types of samples;
— in favourable cases, high sensitivity and precision;
— separation steps need not be quantitative.

Disadvantages of the method are:

— suitably labelled compounds may be unavailable or expensive;
— the tracer must be equilibrated with the sample; there must be complete isotopic exchange;
— the compound isolated for counting and determination of amount (e.g. by weighing) must be chemically pure;
— enough compound must be isolated for a suitable measurement of the amount of material (e.g. mass) to be made.

Perhaps the main disadvantage of the classical method is the necessity of a second type of measurement, so that specific activities can be determined. This limits the sensitivity of the method.

It is beyond the scope of this book to give a comprehensive survey of all the applications of classical IDA in environmental studies. Table 5.1 presents a concise summary, and serves to illustrate the broad range of

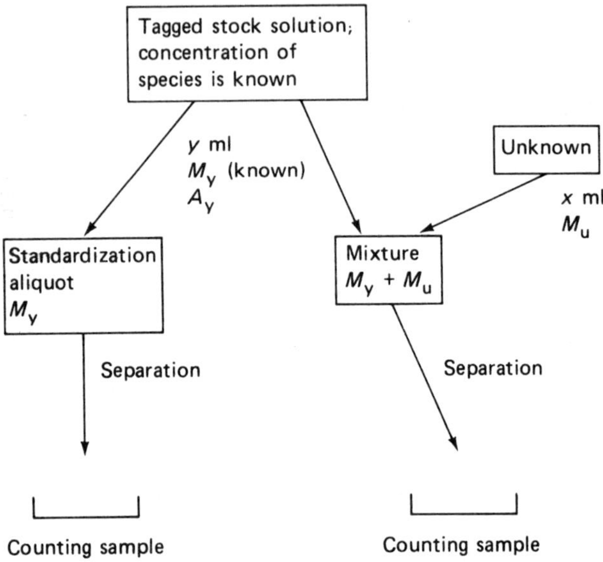

R_s; M_s; $S_s = \dfrac{R_s/E_s}{M_s} = A_s/M_s$

R = cpm
M = mass
E = detection efficiency
A = activity
S = specific activity = A/M
V = volume

R_x; $S_x = \dfrac{R_x/E_x}{M_x} = A_x/M_x$

$$A_y = S_yM_y = S_x(M_y + M_u)\ ;\ M_u = M_y\left[\frac{S_y}{S_x} - 1\right]$$

If $E_x = E_s$, then:

$$M_u = M_y\left[\frac{R_s/M_s}{R_x/M_x} - 1\right]$$

Fig. 5.2 —Classical isotope dilution analysis.

elements and sample types that have been studied. It includes at least thirteen elements, and sample matrices such as water, soil, plants and animals. Separation procedures include precipitation, distillation, solvent extraction and co-precipitation. Methods used for yield (mass) determinations include gravimetry, colorimetry, fluorimetry and flame photometry [7].

5.3 SUBSTOICHIOMETRIC IDA

As pointed out above, the principal handicap of classical IDA is the need for measurement of the amount of pure substance isolated, so that specific activities can be determined. This requirement limits the sensitivity of the method, which is important in the analysis of environmental samples.

However, in the basic equation, only the ratio of specific activities occurs. This means that if exactly the same amount (or an exact multiple of it) of material can be isolated from both the sample and the standards, the equations are greatly simplified. This concept is the basis of a very important branch of IDA, namely substoichiometric IDA, which is outlined in Fig. 5.3.

Table 5.1 — Determination of trace elements in environmental samples by classical IDA

Element	Matrix	Tracer	Separation process	Mass determination	Amount determined
Ba	Binding materials	^{140}Ba	Sedimentation		0.23–1.3 mg
	Biological material	^{140}Ba	Precipitation	Gravimetry	
Ca	Binding materials	^{45}Ca	Sedimentation		1.2 mg
	Soils (exchangeable Ca)	^{45}Ca			
	Water hardness	^{45}Ca	Precipitation	Gravimetry	
Ce	Vegetables	^{130}Ce	Precipitation	Gravimetry	
Cl	Biological fluids and tissues	^{36}Cl	Distillation	Amperometric titration	20–40 meq
	Human body	^{82}Br	Precipitation		
Co	Biological materials, Vitamin B_{12}	^{60}Co	Solvent extraction	Spectrophotometry	4–150 μg
Cr	Marine orgamisms	^{51}Cr	Solvent extraction precipitation	Colorimetry	1.0–5.0 μg
Cs	Sea-water	^{137}Cs	Co-precipitation		
I	Biological material	^{131}I	Distillation	Colorimetry	10 μg
Na	Exchangeable Na in the human body	24			
P	Soil	^{32}P	Precipitation	Gravimetry	
Se	Plant material	^{75}Se	Solvent extraction	Fluorimetry	0.2–6.6 ppm
Sr	Binding materials	^{89}Sr	Sedimentation		mg
	Sea-water	^{85}Sr	Co-precipitation		
	Sea-water	^{88}Sr	Co-precipitation	Flame photometry	9–50 ppm
	Soil	^{90}Sr	Precipitation	Gravimetry	
Tl	Chamber dusts	^{204}Tl	Solvent extraction	Colorimetry	30–300 μg
U	Sea-water	^{237}U	Co-crystallization	Fluorimetry	
	Sea-water	^{237}U	Solvent extraction	Fluorimetry	3.0 μg

Thus, in principle at least, it is only necessary to isolate exactly equal (or exactly multiple) amounts of a desired compound from both sample and standard. The analyst can fix the amount isolated by adding a controlled, substoichiometric, amount of suitable reagent to the sample and standard prior to any separation steps, hence the name 'substoichiometric IDA'. The exact amount of reagent consumed, and the exact ratio in which the analyte reacts with it need not be known; only a constant ratio is necessary. Of course, the reaction product must be separable from the sample and standard, and must also be suitable for counting.

Another important advantage of substoichiometric IDA is enhanced selectivity due to competition for the separation reagent between the analyte and any interfering substances. In favourable cases, this competition will minimize the transfer of interferents into the final sample for counting. Selectivity can be further enhanced by using masking agents to decrease the effective concentration of the interferents, or by using a preconcentration step.

It is clear that the selection of a suitable separation reagent is a critical step in the development of a substoichiometric method. Some criteria include:

— sufficiently large stability constants for any complexes or other separable compounds formed, and significant differences between these constants for different elements (selectivity);
— stability (thermal, photochemical) in solutions at extremely low concentrations;
— formation of only one type of complex (or compound) with the analyte under a given set of conditions;
— favourable kinetics of complex (or compound) formation and of separation;
— no sorption on reaction container surfaces.

Some of the methods used for isolating equal amounts of material from sample and standard are:

— extraction of complexes;
— electrolysis;
— sorption;
— ion-exchange reactions;
— precipitation.

There has been a considerable effort to expand the scope of substoichiometric IDA and several variations have been proposed [7]:

— direct substiochiometric IDA, with the addition of a labelled element, or with induced radioactivity;
— reverse substoichiometric IDA, including activation analysis with sub-

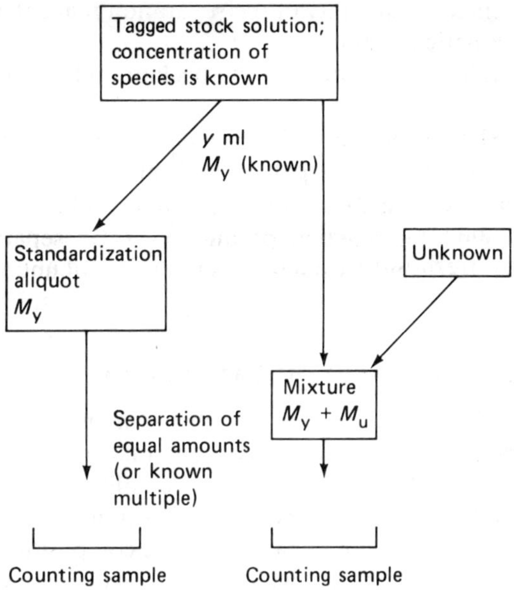

$$S_s = \frac{R_s/E_s}{M_s} = A_s/M_s \qquad S_x = \frac{R_x/E_x}{M_x} = A_x/M_x$$

$M_s = M_x$ or $M_s = a\,M_x$: Equal amounts, or an exact multiple

$$M_u = M_y\left[\frac{R_s/M_s}{R_x M_x} - 1\right]$$

$$M_u = M_y\left[\frac{R_s}{R_x} - 1\right] \text{ or } M_u = M_y\left[\frac{a\,R_s}{R_x} - 1\right]$$

Fig. 5.3 — Basic substoichiometric IDA

stoichiometric separation and determination of the concentration of naturally occurring radioactive elements;

— displacement substoichiometry, in which the basic reaction is not the formation of a stable complex with the analyte, but the partial decomposition of a complex by fractional displacement with a second substance which has a higher stability constant;

— redox substoichiometry, in which partial oxidation or reduction of the analyte is followed by the transfer of the reacted (or unreacted) portion into a second phase.

The sensitivity of substoichiometric IDA is usually limited by one of the following factors:

— specific activity of the tracer;

— chemical stability of the separation reagent and separation product;
— kinetic factors;
— values of any blanks, and background count-rates.

Substoichiometric IDA techniques have been developed for an impressive number of elements in environmental samples. A partial, but illustrative, list of applications is given in Table 5.2. Seventeen elements are included, measured in samples as diverse as biological materials, rocks and ores, fresh-water, sediments, blood, teeth and sea-water.

5.4 SUB-SUPEREQUIVALENCE IDA

The basic requirement of substoichiometric IDA poses a stringent constraint, namely, that equal amounts of material must be isolated from sample and standard. This can be difficult, if not impossible, for a series of samples of widely varying composition or concentration. Ideally, sample and standard should have the same composition.

The sub-superequivalence methods were developed in order to minimize this constraint. The basic method involves the procedure as outlined in Fig. 5.4 [31,32].

The solution to be analysed is labelled with an appropriate isotope, and two series of aliquots are taken. In the first series (Series 1) each aliquot contains the same amount of analyte (x) and activity (A_0). In the second series (Series 2), each aliquot contains K times the amount of analyte and activity used in the first series (Kx and KA_0). The first series is isotopically diluted by adding incremental amounts of the non-radioactive form of analyte ($jy = y, 2y \dots jy$). This gives a series of isotopically diluted aliquots ($x + y, x + 2y \dots x + jy$). If all aliquots of both series are now brought to the same volume, the concentration of one member (actual or hypothetical) in Series 1 will be equal to that of one in Series 2:

$$\frac{x + y_{\mathrm{k}}}{V} = \frac{Kx}{V} .$$

If to all aliquots we add the same amount of a reagent which will react with the analyte, the amount of the product will generally depend on the concentration of analyte in the aliquot. Furthermore, solutions of the same analyte concentration would form the same amount of product. This fact makes the analysis possible, provided that a reproducible separation and counting procedure can be found.

The same amount of product (M_2) is isolated from each of the non-diluted aliquots (M_2 is a constant for Series 2) and each has the same activity (A_2). The aliquots of Series 1 will yield different amounts of product ($M_1 = M_{x+y}, M_{x+2y} \dots$) and activities ($A_1 = A_{x+y}, A_{x+2y} \dots$).

In dilution with non-radioactive materials, the amount of radioactive material remains constant; hence we may write the set of equations given in Fig. 5.4.

Table 5.2 — Determination of trace elements in environmental samples by substoichiometric IDA

Species	Matrix	Tracer	Separation process	Reagent	Amount determined	Reference
Ag	Plant materials	^{110m}Ag	Solvent extraction	dithizone	0.15–0.9 μg/g	[8]
Au	Rock, ores	^{198}Au	Solvent extraction	dioctyl sulphide	2.6×10^{-6}%	[9]
	Rocks	^{198}Au	Solvent extraction	*N*-diethyldithiophosphate in CCl_4	$n\times10^{-7}$%	[10]
Br, Cl	Meteorites, basalts	^{82}Br, ^{36}Cl	Precipitation, dialysis, electrodialysis	$AgNO_3$	10^{-6}–10^{-5} g	[11]
Cd	Biological materials	^{109}Cd	Solvent extraction	dithizone	0.16–0.88 μg/g	[7,12]
Cu	Blood	^{64}Cu	Solvent extraction	NaDDC	$n\times10^{-3}$%	[13]
F	Teeth (enamel)	^{18}F	Adsorption	glass as sorbent	100 μg/g	[14]
	Fluorinated, non-fluorinated water, urine	^{18}F	Solvent extraction	$(CH_3)_3SiCl$ in benzene	0.46–1.03 μg/g	[15]
	Plant materials	^{18}F	Solvent extraction	$(CH_3)_3SiCl$ in benzene	0.05 μg/g	[16]
Fe	Biological materials	^{59}Fe	Solvent extraction	8-quinolinol	550 ppm	[17]
	Rock	^{59}Fe	Ion-exchange	8-quinolinol EDTA–cation exchange column	1.9–9.32%	[18]
Hg	Biological materials	^{203}Hg	Solvent extraction	dithizone	50–70 μg	[19]
	Mineral water	^{203}Hg	Solvent extraction	dithizone	μg/ml	[3]
	Plants (rice)	^{203}Hg	Solvent extraction	dithizone	μg/g	[20]
I	Environmental samples	^{131}I	Dialysis	$AgNO_3$	10^{-7}–10^{-2} g	[21]
Mn	Biological materials	^{54}Mn	Synergistic extraction	TTA+1,10-phenanthroline	68–235 μg/g	[22]
Pb	Granites,	ThB	Solvent extraction	dithizone	28–40 μg/g	[23]
	Rock		Solvent extraction	EDTA–dithizone	7.3–56.2 μg/g	[24]
SO_4^{2-}	Atmospheric precipitates	^{35}S	Precipitation	$BaSO_4$/EDTA	40 ng/ml	[25,26]
Sr	Biological materials	^{90}Sr	Solvent extraction	8-quinolinol	50–70 μg	[27]
Tl	Minerals, sediments, hydrothermal waters	^{202}Tl	Solvent extraction substoichiometric displacement	thallium (I) dithizonate in CCl_4	42 ppm–0.88%	[28]
U	Granites	^{233}U	Solvent extraction with TTA in $CHCl_3$	Arsenazo III	1–40 ng	[29]
Zn	Sea-water	^{65}Zn	Solvent extraction	dithizone	68 μg/l.	[30]
	River water	^{65}Zn	Ion-exchange	8-quinolinol-5-sulphonic acid, anion-exchange column	0.12 μg/l.	[31]

Step 1. Preparation of two series of aliquots from tagged sample.

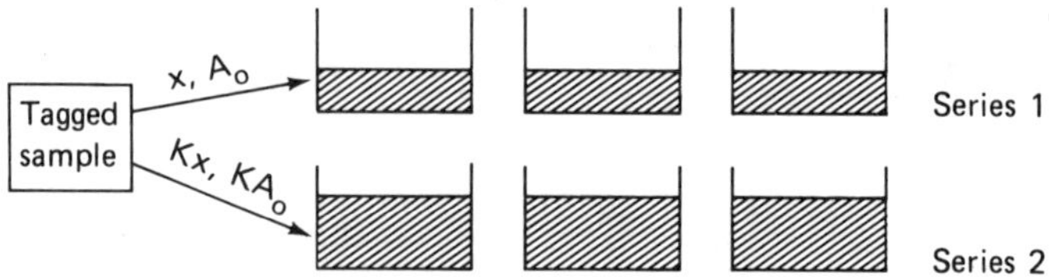

Step 2. Addition of incremental amount of non-radioactive form of analyte to series 1.

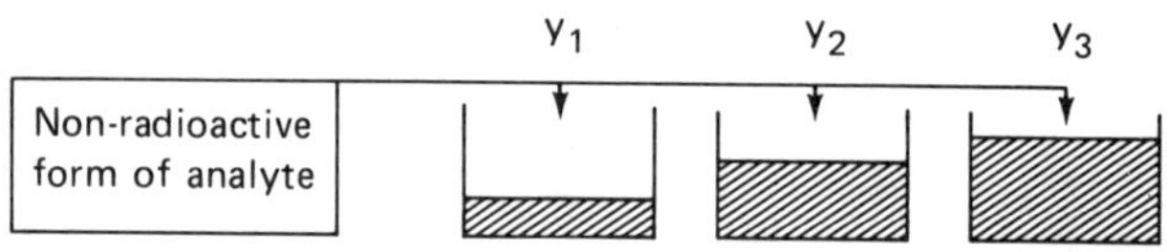

Step 3. Adjustment of all aliquots to same volume.
Step 4. Addition of same amount of separation reagent to all aliquots. Formation of separation product.
Step 5. Separation and counting of separation product

Conservation of activity: $(x+jy)A_{x+jy}/M_{x+jy} = Kx/K\,(A_2/M_2) = xA_2/M_2$

$$I_j = A_2/A_{x+jy} = \frac{M_2}{M_{x+jy}}\left[\frac{x+jy}{x}\right] = \frac{M_2}{M_{x+jy}}\,[1+jy/x]$$

In principle we can make a solution in series 1 such that

$$M_{x+jy} = M_{kx} = M_2$$

For this solution

$$A_2/A_{x+jy} = K$$

The unknown amount of material, x, can be found by plotting I_j against jy, reading the jy value corresponding to $I_j = K$, and applying the equation

$$x = jy/(K-1)\ .$$

Fig. 5.4 — Sub-superequivalence IDA

The amount of unknown (x) in a given volume of sample (V) can be determined from a plot of A_2/A_{x+jy} against jy by reading the value of jy that corresponds to $A_2/A_{x+jy} = K$. Note that it is not necessary to measure M_{x+jy} or M_2, which are the amounts of material isolated from Series 1 and 2.

Four variations of the basic method have been proposed, based on increasing the amount of radioactive tracer instead of that of inactive material, and dilution with the radioactive trace instead of with the non-radioactive species.

The main advantages of the sub-superequivalence method include:

— higher sensitivity in systems where stability constants limit the sensitivity of substoichiometric IDA;

— systems can be studied in which changing complex composition makes isolation of equal amounts of material impossible;
— only one phase need be measured.

Sub-superequivalence methods have been used in a growing number of instances for the analysis of environmental samples. Table 5.3 gives a summary of examples. At least eleven metals have been determined, mainly in water samples.

It is apparent that IDA has become a valuable method for the analysis of environmental samples. Figure 5.5 is a periodic chart, showing those elements (26) which have been determined in environmental samples; those (55) which have been measured by classical IDA; elements measured by substoichiometric IDA (41), and those (12) which have been measured by sub-superequivalence IDA.

5.5 ACCURACY, PRECISION AND SENSITIVITY OF IDA

5.5.1 Accuracy

The most common potential causes of systematic (determinate) errors in IDA are as follows:

(1) Incomplete isotopic exchange [41] may be a cause of serious error if the analyte and diluent are in different chemical or physical states. Precautions should be taken to ensure complete exchange, which may include a series of redox or precipitation/dissolution steps. Heterogeneous systems can be particularly troublesome in this respect.
(2) Undesirable isotopic exchange may occur if the radioactive atom in the species of interest can exchange with another species not related to the procedure. If the reagents used in the analysis contain significant amounts of the analyte, or any substances which undergo isotopic exchange with it, the results will probably show an unacceptable error. Therefore, a blank should always be carried through the same procedure, and very often reagents must be purified for accurate results to be obtained.
(3) Erroneous measurements of radioactivity. The best way to minimize this kind of error is to apply sound laboratory practice, such as:

    — constant sample/detector geometry;
    — appropriate corrections for dead-time and self-absorption;
    — decay corrections when necessary.

(4) Differences in molecular weights (atomic weights of isotopes). Since the relative concentration of radioactive isotopes is usually quite low, this effect is generally negligible. This may not be the case when working with stable or long-lived isotopes.
(5) Fractionation of isotopic substances during purification steps. This effect can be serious when there is a significant difference between the

Table 5.3 — Determination of trace elements in environmental samples by sub-superequivalence IDA

Element	Matrix	Tracer	Separation process	Reagent	Amount determined	Reference
Ag	Water	^{110m}Ag	Solvent extraction	dithizone		[34]
Cd	Water	^{109}Cd	Solvent extraction	dithizone	0.9 μg/ml	[35]
Cu	Water	^{64}Cu	Solvent extraction	dithizone		[34]
	Fresh water	^{64}Cu	Solvent extraction	thio-oxine	10^{-7}–10^{-6} g	[36]
Co	Vitamin B_{12} solution	^{60}Co	Ion-exchange	EDTA	0.81 μg/ml	[37]
Fe	Wate	^{59}Fe	Solvent extraction	cupferron		[34]
	Water	^{59}Fe	Solvent extraction	EDTA–acetyl-acetone	0.03 μg/ml	[38]
Hg	Water	^{203}Hg	Solvent extraction	dithizone		[34]
Se	Water	^{75}Se	Solvent extraction	NaDDC–CCl_4	1.92 μg/ml	[39]
Sr	Water	^{89}Sr	Solvent extraction	DC-18-crown-6	1.12 μg/ml	[35]
Tl	Tl solution	^{204}Tl	Solvent extraction	Brilliant Green–benzene	0.5–105 μg	[40]
U	Environmental samples	^{233}U	Solvent extraction	TTA		[34]
Zn	Fresh and sea water	^{65}Zn	Solvent extraction	thio-oxine	10^{-7}–10^{-6} g	[36]

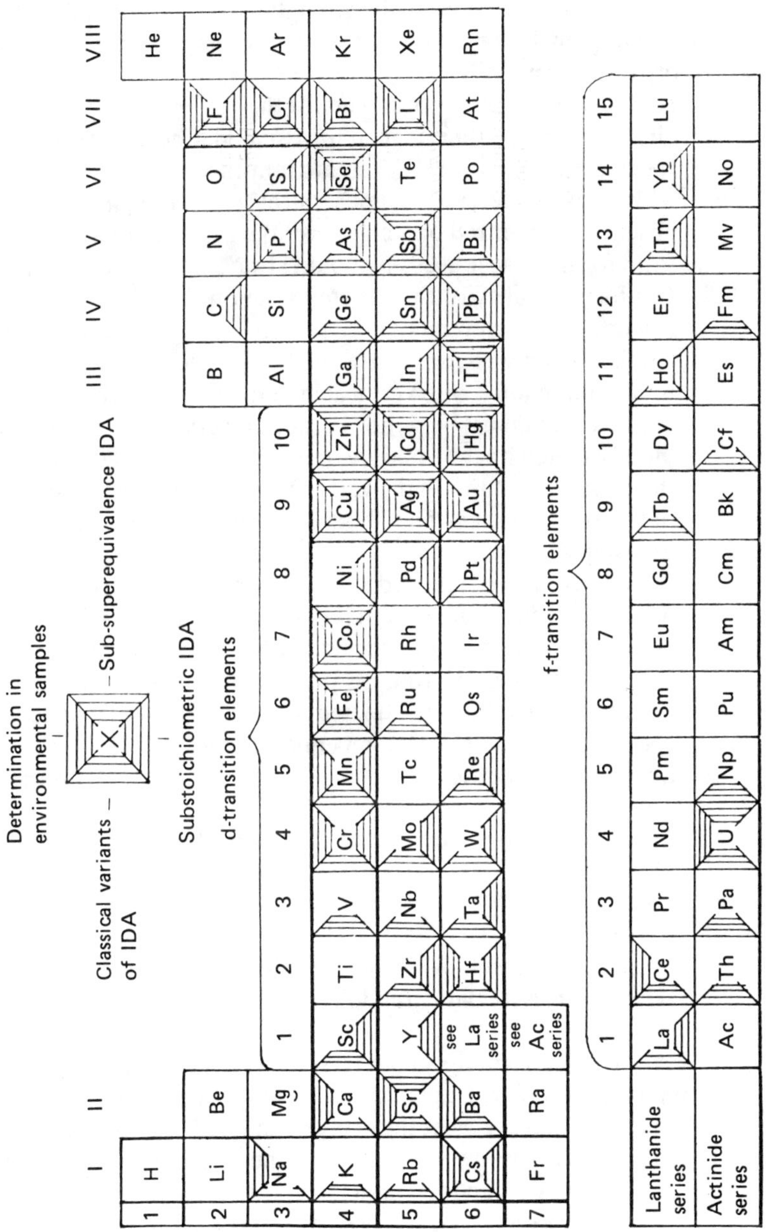
Determination in environmental samples
Classical variants of IDA
Sub-superequivalence IDA
Substoichiometric IDA
d-transition elements
f-transition elements
Lanthanide series
Actinide series

Fig. 5.5.

atomic weights of the isotopes involved (e.g. ^{1}H, ^{2}H, ^{3}H). Low molecular-weight compounds containing these isotopes can become isotopically fractionated during separation processes which are controlled by kinetics or the strength of chemical bonds. This phenomenon can be especially important when working with H-labelled organic compounds.

(6) Radiochemical purity of tracers. If the tracer used contains radioactivity in a form other than that assumed, the observed decrease in specific activity will be different from that corresponding to dilution with pure material. Radiochemical purity requires not only the absence of other radionuclides, but the presence of only one chemical form for the tracer used. It should always be checked for any tracer.

(7) Insufficient purity of the isolated fraction. The purity of the fraction isolated for measurement may be checked by discovering whether the specific activity changes with further purification. The purity of the isolated fraction is often critical in derivative analysis, where excess of radio-reagent must be removed, or in reverse IDA, where the activity of the isolated fraction may be only a few per cent of the total activity of the original mixture. In the variation using two radionuclides, the ratio of the two counting-rates should not change with additional purification steps.

(8) Decomposition of labelled substances can cause errors if it leads to a decrease in the specific activity of the tag. This can be serious when materials of very high specific activity are used. If decomposition occurs after dilution (during purification or isolation), the result is usually not significantly affected, because the labelled and unlabelled species are decomposed to an equal extent.

(9) Superstoichiometric amounts of reagent in a substoichiometric method.

(10) Other systematic errors, such as memory effects in automated systems [42].

5.5.2 Precision

The precision of a result or method will depend on a number of indeterminate errors (non-systematic, random).

(1) Errors in determining amounts isolated, e.g. by weighing, spectrophotometry or polarography. Such errors may be difficult to evaluate experimentally, because they may be coupled with other random errors.

(2) Errors in measuring the activity of the isolated portion. These errors arise from the random nature of the decay process, variations in background levels, instability of counting equipment, irreproducibility of sample/detector geometry and similar factors. In order to minimize this source of error, tracers with high specific activities should be used, isolation and purification yields should not be too low, counting should be done for a sufficiently long time, stable counting equipment should be used and an optimum dilution ratio selected.

(3) Random errors in the amount of tracer or inactive carrier added, which are generally smaller than other errors.
(4) Random contamination by airborne activity, contaminated apparatus, etc., which may be important in trace analysis of environmental samples.

5.5.3 Sensitivity

The sensitivity of IDA is limited by the following factors.

(1) The smallest amount which can be determined or purified in direct IDA.
(2) The original specific activity in reverse IDA.
(3) Specific activity of the diluting radioactive tracer or of the radioactive reagent in derivative IDA.
(4) Equilibrium constants of extraction, hydrolysis, precipitation and similar separation reactions used in substoichiometric IDA.
(5) Blank values (reagent contamination).
(6) Stability of reagents at low concentrations and sorption on surfaces in substoichiometric IDA.
(7) Volumes of solutions used for substoichiometric IDA.
(8) Neutron, photon, or charged-particle flux in IDA after activation.
(9) Interference.

The limit for the specific activity of the tracer is the use of carrier-free radioisotopes, which is seldom possible in practice. Factors (3)–(7) are typical for substoichiometric IDA, with (3) being the most important. It must be realized that this factor includes the counter efficiency and background. With automated procedures, the time available for counting may be the most important factor, and such procedures are generally less sensitive than manual methods.

The finite values of equilibrium constants explain why very low concentrations often cause difficulty in isolating precisely equal amounts of a species from two different systems. Large solubility products or small complex formation constants are good examples.

Blank values, rather than instrumental factors, may limit the sensitivity of substoichiometric IDA. They can be reduced by careful purification of reagents. Several practical suggestions for lowering blank values are given in the monograph by Růžička and Starý [42].

The principle and decisive advantage of IDA and all its variations is the possibility of using non-quantitative isolation procedures; in some cases, this is the only way of doing the required analysis. This possibility allows the analyst to perform an isolation more quickly, choose a purification method from a wide selection, and tolerate a partial decomposition of the species measured.

Moderately irreproducible losses during a procedure can be tolerated in IDA, since total yield is not required. Because the analyte element is radioactive, its path through the analytical scheme can be followed to check identity and purity (through changes in specific activity). Certain types of IDA are suitable for automation.

Except in rare cases, IDA is not a multielement method, and it may not be practical to store radioactive isotopes of the large number of elements, or to prepare and keep the large assortment of labelled compounds required for a multielement programme. Moreover, the availability of suitable isotopes and labelled compounds, particularly organic species, may be a problem.

Some of the major drawbacks of substoichiometric IDA are the need for ultra high-purity reagents, and the very low concentrations at which isolations must be performed. Therefore, IDA may not be suitable for the determination of trace components in environmental samples which require large amounts of concentrated reagents for sample preparation, because it may be difficult to purify these reagents adequately.

A long-term limitation to the use of stable isotopes in IDA is the need for complicated and expensive mass spectrometers.

REFERENCES

[1] G. V. Hevesy and R. Hobbie, *Z. Anal. Chem.*, 1932, **88,** 1.
[2] I. E. Starik, *Problemy sov. geologii,* 1933, **3,** 70.
[3] J. Růžička and J. Starý, *Atomic Energy Rev.*, 1964, **2,** 3.
[4] J. Růžička and J. Starý, *Talanta,* 1961, **8,** 228.
[5] J. Růžička and J. Starý, *Chem. Listy,* 1963, **57,** 1025.
[6] J. Tölgyessy, T. Braun and M. Kyrš, *Isotope Dilution Analysis,* Pergamon Press, Oxford, 1972.
[7] I. Kašparec, A. Zeman and J. Prášilová, *J. Radioanal. Chem.*, 1979, **2,** 281.
[8] N. Suzuki, *J. Chem. Soc. Japan, Pure Chem. Sect.*, 1959, **80,** 370.
[9] E. M. Gilbert, G. G. Glukhov, L. A. Larionova and V. A. Mikhailov, *Izv. SO Akad. Nauk USSR,* 1973, **3,** 85.
[10] E. E. Rakoviskii, B. L. Serebryani and V. V. Severin, in *Analiz i tekhnologya blagorodnykh metallov,* p. 291, Metallurgiya, Moscow, 1971.
[11] I. V. Golubtsov, and V. R. Ignatov, *J. Anal. Chem. USSR,* 1978, **33,** 405.
[12] B. Bibr, J. Lener and A. Zeman, *J. Radioanal. Chem.*, 1969, **3,** 81.
[13] L. A. Kuranova and L. N. Zhuk, in *Yaderno-fizicheskie metody analiza v kontrole okruzhayuschei sredy,* p. 118, Gidrometeoizdat, Leningrad, 1980.
[14] J. N. Kudahl, J. H. Fremlin and J. L. Hardwick, *Proc. Conf. Radioisotopes in Phys. Sci. and Ind.*, Vol. 2, p. 317, IAEA, Vienna, 1965.
[15] H. A. Das, W. H. Köhnemann, G. D. Wals and J. Zonderhuis, *J. Radioanal. Chem.*, 1975, **25,** 261.
[16] G. D. Wals and H. A. Das, *Radiochem. Radioanal. Lett.*, 1976, **26,** 353.
[17] N. Suzuki, S. Nakamura and H. Imura, *J. Radional. Chem.*, 1980, **57,** 37.
[18] E. Gundersen and E. Steinnes, *Talanta,* 1976, **26,** 44.
[19] I. E. Zimakov and T. S. Bogdanova, *Tr. Vses. Nauchno-Issled. Inst., Vet. Sanit.*, 1974, **48,** 147.
[20] N. Suzuki, *Sci. Repts. Tohoku Univ., Ser. 1,* 1959, **43,** 161.
[21] J. M. Lo, J. C. Wei and S. J. Yeh, *Anal. Chim. Acta,* 1977, **93,** 301.
[22] N. Suzuki, S. Nakamura and H. Imua, *J. Radioanal. Nucl. Chem.*, 1984, **81,** 37.
[23] S. M. Grashchenko and E. V. Sobotovich, in *Radiokhimicheskie metody opredeleniya mikroelementove,* p. 73, Nauka, Moscow, 1965.
[24] P. Aruscavage, *Anal. Chim. Acta,* 1976, **82,** 343.
[25] D. Klockow, H. Denzinger and G. Ronicke, *Chemie-Ingenieur-Technik,* 1974, **46,** 831.
[26] W. J. Mayer, D. C. Lechman and D. C. Beyerlein, *Intern. J. Appl. Radiat. Isotopes,* 1975, **26,** 233.
[27] I. E. Zimakov and V. V. Usatenko, *Tr. Vses. Nauchno-Issled. Vet. Sanit.*, 1974, **48,** 144.
[28] A. Adámek and M. N. Chiriatev, *J. Radioanal. Chem.*, 1975, **26,** 51.

[29] S. M. Grashchenko and E. V. Sobotovich, *Radiokhimiya,* 1967, **9,** 412.
[30] Yu. M. Petrov, *Zh. Analit. Khim.*, 1974, **29,** 686.
[31] H. Akaiwa, H. Kawamoto and K. Ogura, *Talanta,* 1977, **24,** 394.
[32] J. Klas, J. Tölgyessy, J. Lesný and E. H. Klehr, *Isotopenpraxis,* 1978, **14,** No. 2, 49.
[33] J. Klas, J. Tölgyessy and J. Lesný, *Sub-super ekvivalentová izotopová zrieďovacia analyzia,* Veda, Bratislava, 1985.
[34] J. Lesný, J. Tölgyessy, Z. Zacharová and O. Rohoň, *Hydrochémia '80,* p. 119, ČSVTS, Bratislava, 1980.
[35] Z. Koreňová, *Thesis,* Slov. Techn. Univ., Bratislava, 1986.
[36] E. I. Grosheva, Yu. A. Bankovskii, O. E. Ververis and A. V. Sturis, in *Yaderno-fizicheskie metody analiza v kontrole okruzhayushchei sredy,* p. 117, Gidrometeoizdat, Leningrad, 1980.
[37] J. Lesný, J. Tölgyessy and J. Klas, *Radiochem. Radioanal. Lett.*, 1976, **26,** 363.
[38] M. P. Chacharkar, J. Lesný, J. Tölgyessy and J. Klas, *Radiochem. Radioanal. Lett.*, 1978, **34,** 399.
[39] J. Lesný, J. Tölgyessy and E. H. Klehr, *Radiochem. Radioanal. Let.*, 1977, **34,** 77.
[40] V. R. S. Rao, Ch. Rulla Rao and G. Tataiah, *Radiochem. Radioanal. Lett.*, 1977, **29,** 261.
[41] A. K. Lavrukhina, T. V. Malysheva and F. I. Pavlotskaya, *Chemical Analysis of Radioactive Materials,* Iliffe, London, 1967.
[42] J. Růžička and J. Starý, *Substoichiometry in Radiochemical Analysis,* Pergamon Press, Oxford, 1968.

6

Radio-reagent methods

Radio-reagent methods are based on the reaction of the substance to be determined, with a suitable reagent, which is usually labelled. The amount of radioactive reagent consumed is determined after completion of the reaction (or after a certain time) by measuring the radioactivity of one of the phases of the system. In some cases the fraction of the substance consumed in the reaction is determined, not by the radioactivity of the reagent, but by measuring the radioactivity of the analyte or of some substance which is able to compete (exchange) with the analyte in its reaction with the reagent, (which in this case is usually non-radioactive) [1–3].

The classification of radio-reagent methods is a difficult problem [3]. According to the type of chemical interaction and relationship of the analyte to the radioactive substances measured, radio-reagent methods can be divided into three basic groups [4]:

(1) the radioactive substance is a typical reagent which is able to react with the compound to be determined;
(2) the method is based on a "non-equivalent" competition; the radioactive substance is able to exchange with the analyte;
(3) the method is based on an "equivalent" competition, *i.e.* the labelled substance is chemically identical with that to be determined.

The difference between groups depends on which of the substances involved is radioactive. For example, if the precipitation of barium or radium sulphate is used in a determination, the following variants may occur: in the first the inactive barium is determined by precipitation with an excess of radioactive sulphate (or inactive sulphate is determined by a surplus of radioactive barium); in the second the barium is determined by precipitation with an insufficient amount of sulphate in the presence of radium which acts as an indicator, so the higher the barium content, the lower the fraction of barium precipitated and, correspondingly, of radium co-precipitated; in the third a constant but insufficient amount of sulphate can be used for the determination of barium, the indicator for precipitation completion being radioactive barium.

From the practical point of view, it is convenient to divide radio-reagent methods into three groups: simple radio-reagent methods, methods of concentration-dependent distribution and radio-release methods. For the analysis of environmental samples, radio-release methods are widely used.

6.1 SIMPLE RADIO-REAGENT METHODS

These methods utilize a reaction with a suitable reagent, and sometimes with a third substance; one of the reagents is radioactive and the reaction is quantitative.

Simple radio-reagent methods can be classified into the following three groups:

(1) determination with labelled reagents;
(2) determination with labelled analyte;
(2) determination based on isotope exchange.

6.1.1 Determination with labelled reagents

This fundamental and most widely used type of simple radio-reagent methods is based on a quantitative reaction occurring between the non-radioactive analyte and a labelled reagent. Usually an excess of radioactive reagent solution of known analytical concentration is used. However, it is necessary to separate the active product from the excess of radioactive reagent. For this reason, this procedure makes use of various separation techniques, such as precipitation, formation of extractable chelates or sorption. The majority of determinations are based on classical precipitation reactions.

The principle of determination is the addition of an excess of radioactive precipitant (*B) to a non-radioactive analyte (A). Excess of reagent (*B) is separated from the precipitate and the radioactivity of the latter is measured either directly, or indirectly [5] from the known radioactivity added and the radioactivity of the supernatant liquid. If the specific activity of the reagent (cpm/mole) and the composition of the precipitate are known, the quantity of analyte can be calculated. Therefore measurement of radioactivity, and not weighing, is used in the determination.

In some cases, the activity of the precipitate itself is measured instead of that of the supernatant liquid. The drawbacks of this procedure are increased complexity, the need to wash the precipitate, possible losses due to dissolution during washing, an increase of the effect of self-absorption during measurement of the activity, and lower accuracy in cases where the major part of the radioactive reagent is consumed in forming the precipitate. This procedure may be advantageous when circumstances require a considerable excess of radioactive precipitant, or if the precipitate adsorbs part of the excess of reagent and the latter can be conveniently removed by washing.

A portable apparatus has been developed for the determination of chloride in water (0.001%) by precipitation with silver labelled with ^{110m}Ag.

Traces of anionic surface-active substances in river and drinking waters are determined by extraction of their complexes with $^{59}Fe(II)$-(1,10-phenanthroline)$_3^{2+}$ into chloroform. A rapid and routine method for the determination of sulphate in soils is based on its precipitation with a barium salt labelled with ^{133}Ba. The ^{131}Ba nuclide has been used for the determination of sulphur in small samples of animal tissue. Phosphate content in fertilizers and minerals can be determined by precipitation with ammonium molybdate (^{99}Mo) [3].

In some cases filtration, sedimentation, or centrifugation can be omitted, for example when the radioactive precipitant is a gas. ^{14}CO has been used in this way to determine haemoglobin in blood [6].

There are several determinations in which the precipitate with the radioactive reagent is not produced by the original form of the analyte, which has to be converted into a compound which will form a precipitate. Thus, for example, precipitation of the complex $Cu(en)_2Hg^{131}I_4$ can be utilized for determining mercury in environmental samples, but the mercury must first be converted into its iodo-complex [7]. Copper can be determined after transformation into its ethylenediamine complex [8].

In some cases, precipitation reactions can be used for radiometric determination even if the exact composition of the precipitate is unknown, or if the reagent used forms both a precipitate and a colloidal solution. The procedure in such cases is to construct a purely empirical calibration curve, or to use the method of known additions. For example, proteins have been determined on the basis of their reaction with heteropolytungstate (^{185}W) to form complexes that can be collected on membrane filters. The radioactivity of the collected complex is directly proportional to the original quantity of albumin.

Excess of radioactive reagent can, of course, be separated by means other than precipitation. For example, cadmium can be determined by the extraction of $Cd(pyridine)_2I_2$ into benzene. Radioactive $K^{131}I$ is used and the activity of the organic phase is proportional to the quantity of cadmium. To prevent the oxidation of iodide to iodine, ascorbic acid is added [9].

6.1.2 Determination with labelled analyte

This type of radio-reagent method is the opposite of determinations using a labelled reagent. Both groups of determinations have a common feature in that the radioactively-labelled substance must be present in excess. Should the non-radioactive component be in excess, then all the added radioactivity, irrespective of the quantity of the components, would be found in the phase of the reaction products and a determination would be impossible.

In this method, the radioactive indicator is added to the unknown sample and isotope exchange is allowed to take place; then the reagent is added in a substoichiometric but known quantity. After the reaction has taken place, the phases are separated and the activity is measured in one of them (or sometimes in both) and the quantity of analyte calculated. Besides keeping in mind the low specific activity of the added radionuclide, it should be remembered that a further non-radioactive carrier is being added to the unknown solution. In those cases where the activity of the equilibrium solution is measured, once the reaction is completed the unknown concentration can be calculated from the equation:

$$x+x_0=\frac{A_{\text{init}}\,B}{(A_{\text{init}}-A_{\text{equilib}})}$$

where x is the quantity of analyte (equivalents), x_0 the amount of added radioactive indicator (equivalents), B the amount of unlabelled reagent (equivalents), and A_{init} and A_{equilib} represent the total activity of the initial and equilibrium solutions, respectively.

An example of the use of this method is the determination of nickel with dimethylglyoxime. The nickel, labelled with ^{63}Ni, is precipitated in ammoniacal medium with a substoichiometric quantity of dimethylglyoxime. After centrifugation, the activity of the precipitate is measured and the quantity of nickel calculated.

6.1.3 Determinations based on isotope exchange

This type of determination is based on measurement of the fraction of the radionuclide in one form, and then in a second form which differs either by being a different chemical compound or containing the analyte in a different oxidation state [2, 3, 10, 11]. If complete isotope exchange is reached, the distribution of masses and radioactivities is the same, i.e.

$$\frac{x}{y}=\frac{A_1}{A_2}$$

where x, y are the masses of the element in the two different forms or phases and A_1, A_2 are the respective radioactivities (cpm).

From the known value of y and measurement of $A_1+A_2=A$, the relationship can be modified to:

$$x=\frac{A_1y}{(A-A_1)}$$

and if $y \gg x$ (so that $A_2 \gg A_1$ and, of course $A \gg A_1$) we obtain

$$x=\frac{A_1y}{A}$$

In a calibration procedure the values y and A are constant, so a calibration curve in the form $x=f(A_1)$ is linear.

The prerequisites for successful application of this type of determination are:

(1) a significantly higher concentration of the non-radioactive form in the second phase than that of the radionuclide to be separated;

(2) sufficiently rapid isotope exchange (the average time required for separation by solvent extraction is 5–10 min, by ion-exchange chromatography 10–15 min, and by precipitation 20–30 min);
(3) the absence of other chemical interactions;
(4) an easy transformation of the diluted form into a sample suitable for the measurement of radioactivity.

The condition of separate measurement of radioactivity of the two forms of the substance in question is automatically fulfilled in heterogeneous isotope exchange. Separation methods other than solvent extraction can seldom be used. In the case of homogeneous isotope exchange the two forms must be separated by a special procedure.

Table 6.1 gives some examples of the use of isotope exchange for the determination of trace substances.

6.2 METHOD OF CONCENTRATION-DEPENDENT DISTRIBUTION

Characteristic features of the method of concentration-dependent distribution (CDD) are as follows [28]:

(1) the analysis is based on the distribution of a radioactive substance between two phases or between several parts of the system used (paper chromatography, electrophoresis);
(2) the ratio of the activities in both phases or parts of the system depends strongly on the concentration of the substance to be determined;
(3) the distribution is governed by the values of the corresponding equilibrium or kinetic constants.

The determination of chloride in water samples by using a radioactive AgCl precipitate can be taken as an example. A water sample with an unknown concentration of chloride is shaken with an excess of the labelled AgCl precipitate and the radioactivity of silver is measured in the equilibrium solution. The higher the chloride concentration the lower the concentration of silver in the solution (owing to the constancy of the AgCl solubility product) [29]. The Cl^- ions determined do not react with the AgCl precipitate, they only suppress its dissolution. The amount of chloride in the precipitate has no direct relationship to the concentration of Cl^- ions in the sample solution.

A second example is the determination of barium based on sorption on an iron(III) hydroxide precipitate [30]. It has been found that the distribution coefficient of barium between $Fe(OH)_3$ and the solution (under constant experimental conditions such as phase ratio, pH, concentration of other substances, temperature) depends on the total concentration of barium in the system. If this dependence is determined experimentally, it can be used for the construction of a calibration curve and the determination of barium. In this case, a known amount of a suitable barium radionuclide is added to the unknown amount of a barium salt, and the adsorption is allowed to take place under standard conditions. From the activities of both phases, and use of the calibration curve, the unknown concentration of

Table 6.1 — The use of isotope exchange for the determination of trace substances [3, 10, 11]

Element	Ion determined	System	Type of exchange	Remarks (concentration range, time of exchange, sample type)	Reference
Ag	Ag^+	Ag^+ in 0.5*M* H_2SO_4–^{110m}AgHDz in CCl_4	Hetero	~2 μg/ml; waste waters	[12]
As	As^{3+}	As^{3+} in 2*M* HCl–^{74}As $(DDC)_3$ in CCl_4	Hetero	0.1–1.0 μg As	[13]
	As^{3+}	As $(DDC)_3$–$^{76}AsI_3$ in CCl_4	Homo	0.1–10 μg	[14]
Bi	Bi^{3+}	Bi $(DDC)_3$–$^{210}BiI_3$ in CCl_4	Homo	sensitivity 10^{-8}g Bi in 7 ml	[15]
Ce	Ce^{3+}	Ce $(HA_2)_3$–^{144}Ce (HA_2) in toluene	Homo	0.01–2 μg Ce; exchange in 5–10 min	[16]
Hg	Hg^{2+}	Hg^{2+} in 0.1*M* H_2SO_4–^{203}Hg $(DDP)_2$ in CCl_4	Hetero		[17]
	Hg^{2+}	Hg^{2+}–C_6H_5 $^{203}Hg^+$ in 0.5*M* H_2SO_4	Homo	minerals	[18]
	CH_3Hg^+	CH_3Hg^+–$^{203}Hg^{2+}$ in 0.5*M* H_2SO_4	Homo	environmental samples	[19] [20]
	$C_6H_5Hg^+$	C_6H_5HgCl–$^{203}HgCl_4^{2-}$ in 3*M* HCl	Homo	sensitivity 1 ng/ml in 5 ml, very selective, biological material	[21]
I	I^-	I^- in H_2O–$^{131}I^1$–(I_2) in C_6H_6	Hetero	~1 ng/ml I^-, natural water	[22]
	I^-	I^- in 0.005*M* H_2SO_4–$^{131}I_2$ in CCl_4	Hetero		[23]
P	PO_4^{3-}	PO_4^{3-} in 0.1*M* H_2SO_4–phosphomolybdate (^{32}P) or tetraphenylarsonium in 1,2-dichlorethane	Hetero	exchange <30 sec	[24]
Pb	Pb^{2+}	Pb tartrate–^{212}Pb $(DDC)_2$ in CCl_4	Hetero	0.01–1 μg Pb in 5 ml, 1–2 min exchange, water samples	[25]
Sb	Sb^{3+}	Sb^{3+} in 0.5*M* H_2SO_4–^{124}Sb-$(DDC)_3$ in CCl_4	Hetero	0.01 μg	[13]
	Sb^{3+}	Sb $(DDC)_3$–$^{124}SbI_3$ in CCl_4	Homo	0.5–20 μg	[26]
Tl	Tl^+	Tl (DDC)–^{204}Tl $(DDC)_3$ in CCl_4	Homo	0.02–10 μg Tl, isotope exchange takes <0.5 hr	[27]

Hetero — heterogeneous exchange, Homo — homogeneous exchange, HDDC — diethyldithiocarbamic acid, HDz — dithizone, HA_2 — di(2-ethylhexyl)phosphoric acid, HDDP — dibutyldithiophosphoric acid.

barium, corresponding to the measured ratio $Ba_{precipitate}/Ba_{solution}$, can be found [3].

In classifying the methods of CDD, it is useful to distinguish two extreme types of determination:, *saturation analysis* and *analysis without saturation*.

In *saturation analysis* the cause of the change in the distribution of the radioactive substance is an increase in the saturation of the reagent by the given substance. In the example mentioned above, the $Fe(OH)_3$ adsorbent is successively saturated with barium ions. The term *saturation analysis* originally appeared in the literature in connection with the determination of biochemically important substances (steroids, hormones, vitamins) and is equivalent to the term *radioimmunoassay*.

In the case of *analysis without saturation*, the cause of the change in the distribution coefficient is the shift of the chemical equilibrium due to increase in the concentration of the substance to be determined, irrespective of the extent of saturation. The concentration of the analyte may differ from that of the reagent by several orders of magnitude. The determination of chloride mentioned above can be considered as an example of *analysis without saturation*.

6.2.1 Saturation analysis

The principle of *saturation analysis* [31] consists in the addition to the analyte (P) of a small amount of radioactively-labelled *P and a certain quantity of substance Q, which gives a specific reaction with P:

$$n\text{P} + m\text{Q} = \text{P}_n\text{Q}_m$$

The reaction product can be transferred into a phase different from that containing the unreacted substance P. If to a number of samples of different initial concentrations of P, the same quantity of Q is added, then the ratio of the quantity of the reaction product P_nQ_m to the unreacted quantity of P will generally be higher for smaller initial quantities of P, and the final concentration of the unconsumed Q will also be higher. The isotopic composition of P and P_nQ_m is, nevertheless, identical, and therefore the ratio of radioactivity of the two phases will be determined by the ratio of the amounts of both substances, and this ratio is a measure of the initial concentration of P.

In this manner vitamin B_{12} (P) was determined by means of its reaction with a specific binding protein (Q). The reaction product was adsorbed on active carbon and the unreacted protein remained in solution. Vitamin B_{12} labelled with ^{60}Co was used in this determination.

Toxic metals which form EDTA complexes have been determined by saturation analysis [32–34]. The proportion of metal complexed is lower for higher metal concentration at a constant EDTA concentration. The measure of the degree of complexation is the distribution of the metal between the solution and an anion- or cation-exchange resin. The sensitivity of these methods in many cases exceeds that of other methods of trace analysis (activation analysis, spectrophotometry).

6.2.2 Analysis without saturation

Analysis without saturation has been used for the determination of an extractant, by using the dependence of the distribution ratio of a suitable carrier-free radioactive metal on the extractant concentration under standard conditions. For example, ^{95}Zr and ^{95}Nb have been used for the determination of di-isoamyl phosphoric acid, ^{65}Zn for dodecylbenzene sulphonic acid, and ^{137}Cs for tetraphenylborate and phosphomolybdate.

The distribution may also be between different organs of a living organism. A tissue containing unknown amounts of thiouracil was fed to rats which subsequently received an injection of ^{131}I. After some time the rats were killed, their thyroid glands were cut out, and their radioactivity was measured. The higher the initial concentration of thiouracil, the higher the radioactivity found.

The number of procedures described that use *analysis without saturation* is limited. These procedures are generally characterized by extremely high theoretical sensitivity but low precision [3].

6.3 RADIO-RELEASE METHODS

Radio-release methods are those analytical procedures in which the non-radioactive test substance is brought into contact with a radioactive reagent, and as a result of an appropriate chemical reaction, a portion of the activity of the radioactive agent will be released, i.e. will pass from the solid to the liquid or gas phase, or from the liquid to the gas phase. A characteristic feature of radio-release methods is that the radioactive substance leaves the solid or the liquid phase without any non-radioactive substance replacing it.

The determination is usually done by using a calibration curve, constructed by plotting the quantity of the released radioactivity, or the rate of release, *vs.* the concentration of the analyte. In some cases it is possible to calculate the amount of the analyte from the amount of the radioactivity released, if the specific activity of the reagent and the stoichiometry of the reaction are known. Radio-release methods in which the radioactive reagent is released into the gaseous phase require rather more complicated instrumentation than that used in the case of simple radio-reagent methods.

Radio-release methods may be classified according to the type of the radioactive reagents employed, such as:

(1) radioactive kryptonates,
(2) radioactive metals,
(3) radioactive salts and other substances.

6.3.1 Radioactive kryptonates

The term "radioactive kryptonate" is used for substances into which atoms or ions of the radionuclide krypton-85 have been incorporated, regardless of the structure of the substance. ^{85}Kr can be incorporated into such a variety of materials that it can be considered as a universal radioactive tracer.

The krypton may be incorporated into solid substances in several ways [35–37].

(1) By diffusion of krypton into solid substances from the surrounding atmosphere. The amount of gas absorbed is a function of the time of exposure, partial pressure and temperature.
(2) By bombardment with accelerated krypton ions, targets are obtained where the average projected range of the label is a function of the properties of the target substance and of the bombarding ion.
(3) By bombarding powdered materials and forming them into tablets,

targets are obtained in which the distribution of the label can be assumed to be macroscopically homogeneous (pseudo-homogeneous distribution).

(4) By diffusion of krypton into solid targets for sufficiently long periods, or sublimation condensation of the target substance in a krypton atmosphere, or by diffusion of krypton into melts and subsequent quenching, targets are obtained with a homogeneous krypton distribution.

The basis of the use of radioactive kryptonates in analytical chemistry is that during the chemical reaction the crystal lattice of the kryptonated carrier is destroyed, the carrier consumed and the radioactive krypton released [38–40]. The determination can be done with a calibration curve or by comparison with a standard.

Radio-release methods with the aid of radioactive kryptonates as analytical reagents are very useful for the analysis of environmental samples [41].

6.3.1.1 Determination of air pollutants

In *ozone* determinations with the radioactive clathrate hydroquinone kryptonate the following reaction is used:

$$[C_6H_4(OH)_2]_3\,[^{85}Kr] + O_3 \rightarrow 3C_6H_4O_2 + 3H_2O + {}^{85}Kr\uparrow$$

The logarithm of the activity of the released ^{85}Kr is linearly proportional to the logarithm of the ozone concentration over a concentration range of 10^{-10}–10^{-6} g of ozone per litre of air (kryptonate specific activity of about 3.7 GBq/g). A higher specific activity proportionately increases the sensitivity [42].

The general requirements for analysers of air-composition in the upper layers of the atmosphere are high sensitivity, low weight, small dimensions and low price. These requirements are met in an ozone analyser which consists of the radioactive kryptonate, a measuring cell with an end-window GM counter and a small pump [43]. The analyser, in a plastic case, is attached to the upper part of a normal weather radiosonde and the detected activity is recorded directly.

Another ozone analyser has been described [44] which incorporates a highly efficient reagent produced by the improved process of adsorbing radioactive hydroquinone kryptonate onto calcium silicate.

For the determination of ozone concentration in a gaseous medium containing both ozone and molecular oxygen, the apparatus consists of two reaction cells, one containing an untreated hydroquinone kryptonate (without calcium silicate and platinum-black-coated catalysts), which responds to oxygen in the gas stream, and another containing a treated or catalytic hydroquinone kryptonate, which responds to both ozone and molecular oxygen. The difference in ^{85}Kr readings in the measuring cells indicates ozone concentration.

Chromatographic papers impregnated with hydroquinone and kryptonated by diffusion have been used for the determination of O_3 [46, 46]. This determination is based on the oxidation of the hydroquinone clathrate of ^{85}Kr by O_3 and measurement of the residual activity after a suitable contact time.

Sheets of paper (20 × 2 cm, Whatman No. 2) were immersed in a concentrated solution of hydroquinone, allowed to dry at room temperature and kryptonated. Air–ozone mixtures were allowed to impinge on the paper in 1.5 cm-diameter circles. These circles were then counted to determine the decrease in radioactivity.

The concentration dependence of the relative residual activity was found to be linear in the O_3 concentration range of 100–400 μg/l. The reproducibility of the method was approximately ±4%.

Sulphur dioxide has also been determined with radioactive hydroquinone kryptonate [42,43,47–49]. The determination is based on the mechanism of double release. In the first stage, sulphur dioxide releases chlorine dioxide, which is a strong oxidizing agent. The chlorine dioxide oxidizes the radioactive kryptonate and gaseous ^{85}Kr is released. The following reactions are involved:

$$SO_2 + 2ClO_3^- \rightarrow 2ClO_2$$

(This equation is an oversimplification, since it was found experimentally that one molecule of sulphur dioxide releases between 4 and 8 molecules of chlorine dioxide).

$$ClO_2 + [C_6H_4(OH)_2]_3\,[^{85}Kr] \rightarrow {}^{85}Kr\uparrow$$

The activity of the released ^{85}Kr is linearly dependent on the concentration of sulphur dioxide, which can be determined down to a level of 0.001 ppm. A diagram of the apparatus for the determination of sulphur dioxide in air is given in Fig. 6.1.

The only gases that should interfere with the operation of this analyser are the oxides of nitrogen and ozone.

Chromatographic papers impregnated with hydroquinone and kryptonated by diffusion have been used for the determination of sulphur dioxide [50]. The activity of the released ^{85}Kr is linearly dependent on the concentration of sulphur dioxide, which can be determined down to a level of 0.001 ppm.

In another method, sulphur dioxide has been determined in air with the use of a radioactive kryptonate of iodine [51, 52]:

$$SO_2 + I_2[^{85}Kr] + 2H_2O \rightarrow H_2SO_4 + 2HI + {}^{85}Kr\uparrow$$

Fluorine and fluorine-containing oxidants can be determined by means of radioactive hydroquinone kryptonate. In the apparatus described by Hommel *et al.* [53], a radioactive kryptonate with a specific activity of 16.65 GBq/g was used.

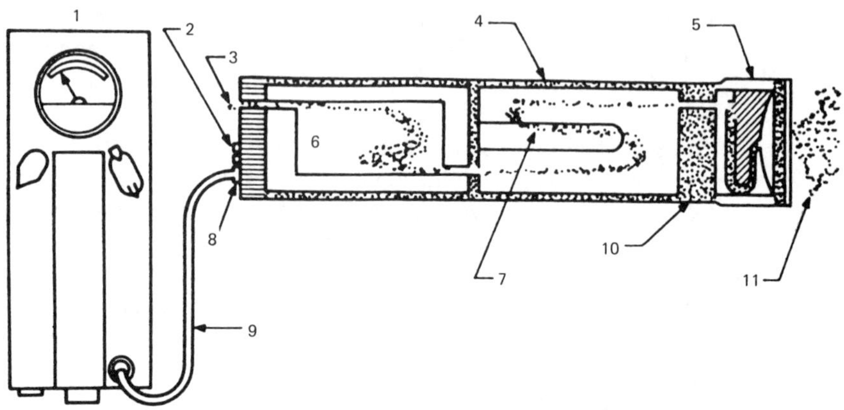

Fig. 6.1 — Diagram of an instrument for the determination of sulphur dioxide: 1 — indicating meter, 2 — power supply, 3 — vent, 4 — detector, 5 — sensing head, 6 — pump, 7 — GM tube, 8 — power and signal connectors, 9 — signal cable, 10 — lead, 11 — air intake.

Jesenák *et al.* [54, 55] suggested a radio-release method for hydrogen fluoride determination based on its reaction with kryptonated silica:

$$SiO_2[^{85}Kr] + 4HF \rightarrow SiF_4 + 2H_2O + {}^{85}Kr \uparrow$$

Disks of monocrystalline silica (13 mm diameter and 1 mm thick) were kryptonated by ion bombardment. A series of hydrogen fluoride solutions was prepared in the molar concentration range 0.0–0.15, which provided equilibrium hydrogen vapour concentrations in the range 0.0–10.0 ppm. Ten-ml portions of the appropriate solutions were equilibrated in 50-ml plastic bottles at 25°C for 2 hr. The kryptonated silica disks were suspended inside the bottles and allowed to react with the hydrogen fluoride vapour. The time dependence of the activity of each disk was determined. A calibration graph for hydrogen fluoride and an exposure time of 6 hr was linear.

The dependence of the residual activity A_r on exposure time t and concentration of hydrogen fluoride (C, ppm) is given approximately by the equation

$$\log A_r = -0.0072\, t \log(1 - 0.0072C).$$

This method may also be suitable for the determination of the average effective hydrogen fluoride concentration over longer exposure times. Hydrogen bromide will not interfere at concentrations up to 150 ppm and hydrogen chloride at 115 ppm decreases A_r by less than 1%.

Mercury vapour can be determined by a radio-release method using kryptonated selenium sulphide [56]. Papers impregnated with a known amount of kryptonated selenium sulphide were kept at 55°C to achieve thermal stability for the kryptonate. They were counted, and a relative specific activity was calculated. In a given experiment, the activity of the sensor was monitored as a function of time for different mercury vapour concentrations. Upon exposure to air containing mercury vapour, the radioactivity of the paper decreased. The decrease in the radioactivity (ΔA), divided by its initial specific activity (S_0), was found to be a linear function of the mercury vapour concentration (c_{Hg}) and of exposure time:

$$-\frac{\Delta A}{S_0} = 7.8 \times 10^{-4} C_{Hg} t$$

where t is in minutes and C_{Hg} in ppm. The excellent linearity implies that the method has good reproducibility.

Various types of apparatus for the determination of oxidizing or reducing atmospheric impurities (F_2, Cl_2, SO_2, NO_2, NO, RNH_2, O_3), in concentrations of the order of ppm, have been developed in the research laboratories of Tracerlab. With these instruments, gases which can react directly or indirectly by chemical, catalytic, or thermal reactions with radioactive kryptonates are measurable [37, 49, 57].

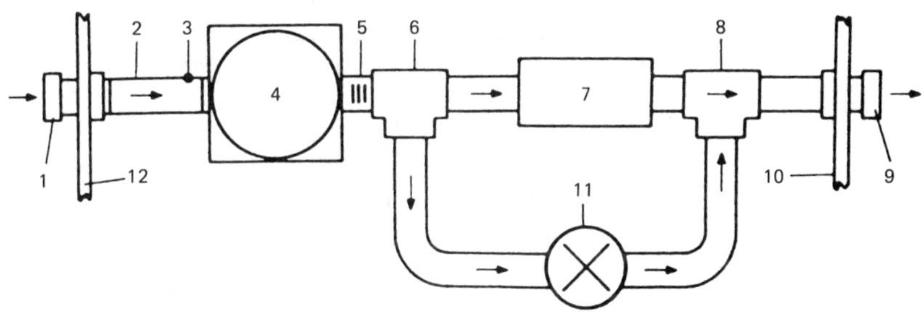

Fig. 6.2 — The flow diagram for an NO_2 monitor: 1 — input, 2 — clathrate cell, 3 — red dot or arrow, 4 — counting chamber, 5 — flowmeter, 6 — T-piece, 7 — pump, 8 — T-piece, 9 — exhaust, 10 — panel, 11 — flow control, 12 — panel.

The flow diagram for a nitrogen dioxide monitor is shown in Fig. 6.2. The basic system contains a sampling pump, flow-valve, counting chamber and radioactive kryptonate (clathrate). Chemical and catalytic cells are included when necessary. The sample gas (air) is pulled through the system by the pump. As the gas of interest enters the instrument, it immediately comes into contact with the chemical and catalytic cells if required, and then reacts with the radioactive kryptonate. The ^{85}Kr released enters a counting chamber of known volume, where its concentration per unit volume is determined. This concentration can be related directly to the volume concentration of the gas of interest entering the system.

A lower detection limit of 0.5 ppm has been achieved. The upper limit of the instrument is dependent on the required cell life and the saturation characteristics of the GM tube.

A schematic drawing of an instrument for the detection of three different gases (CO, NO_x and hydrocarbon vapours from automobile exhausts) is shown in Fig. 6.3 [58]. The sensing substrates are $PtO_2[^{85}Kr]$ for determination of hydrocarbon vapours (detection limit for hexane ~16 ppm) $[C_6H_4(OH)_2]_3[^{85}Kr]$ for determination of NO_x (minimum detectable concentration ~2 ppm) and $PdCl_2[^{85}Kr]$ for determination of CO (minimum detectable concentration ~125 ppm).

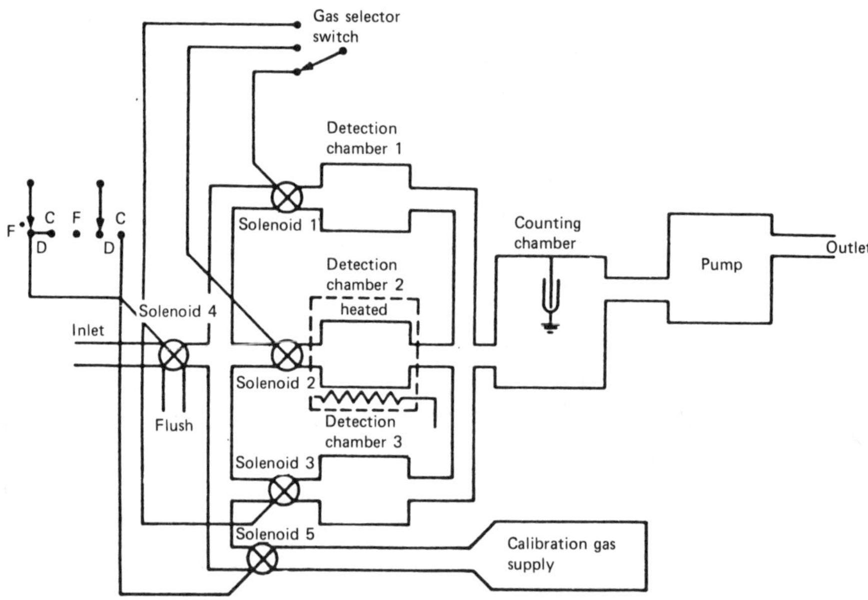

Fig. 6.3 — Schematic drawing of an instrument operating in the ^{85}Kr effluent mode for the detection of three different gases.

6.3.1.2 *Determination of water pollutants*

Dissolved oxygen is a constituent of major concern in water quality studies and pollution control. It is vital in the protection of fish and other important aquatic species which must have adequate oxygen levels in order to survive. Also, the presence or absence of oxygen is a major determinant of water quality. In its presence, water is an oxidizing environment, and hence most inorganic species tend to be in higher oxidation states [Fe(III), Mn(IV), S(VI), N(V), etc.]. In its absence, these same elements tend to be present in reduced forms [*e.g.* Fe(II), Mn(II), S(−II), N(−III), etc.]. Aerobic biological degradation of organic material is relatively rapid, and leads to the formation of carbon dioxide and water. The absence of oxygen destroys the aerobic organisms, and anaerobes take their place. Anaerobic biological decomposition of organic material is relatively slow and energy inefficient, and leaves partially degraded, noxious organic compounds such as acids, alcohols, mercaptans, etc. [59].

Dissolved oxygen can be measured by the use of radioactive thallium kryptonate. The measurement is based on the reaction:

$$4Tl[^{85}Kr] + O_2 + 2H_2O \rightarrow 4Tl^+ + 4OH^- + {}^{85}Kr\uparrow$$

The reaction is stoichiometric, therefore the amount of radioactive krypton released and the reduction of residual activity in the kryptonated thallium are proportional to the dissolved oxygen concentration in the sample, under conditions of constant pH. The thallium carrier can be prepared by electrolytically coating a copper or platinum plate, or copper turnings, with thallium. The carrier is then kryptonated by bombardment with ^{85}Kr ions or by diffusion.

The radioactive thallium kryptonate can be placed in the sample, or the sample is passed through a column of the turnings. Either the released ^{85}Kr or the decrease of activity of the thallium kryptonate can be measured. The analysis can be automated.

For distilled water samples, the decrease in activity of the thallium kryptonate was found to be linearly related to oxygen concentration down to the lowest level investigated (0.3 ppm).

The pH of a water sample can be measured by making use of the fact that the dissolution of thallium kryptonate (and therefore also the release of ^{85}Kr) is pH-dependent under conditions of constant temperature and pressure. These conditions are of course related to oxygen concentration.

To make a pH measurement, the thallium kryptonate disk was submerged in 20 ml of solution for an arbitrary time. The disk was then dried in a desiccator under reduced pressure, and its residual activity measured. The pH was determined from an appropriate calibration chart.

The optimum sample–kryptonate contact time was chosen as 60 sec. The residual activity of the kryptonate was found to be linearly related to the pH under constant experimental conditions of reaction time, temperature and pressure [61–65].

A radio-release method for the determination of *dichromate* makes use of radioactive silver kryptonate according to the reaction:

$$Cr_2O_7^{2-} + 6Ag[^{85}Kr] + 14H^+ \rightarrow 6Ag^+ + 2Cr^{3+} + 7H_2O + {}^{85}Kr\uparrow$$

The electrochemical potentials of the Ag/Ag^+ and $Cr^{3+}/Cr_2O_7^{2-}$ couples are such that the reaction is expected to be quantitative.

The method was developed by using a neutral solution of potassium dichromate in the concentration range 0.0125–0.1*M*. It has been shown that Fe(III), chloride and nitrate interfere with the determination [66].

Vanadate can also be measured in aqueous solution, by using silver kryptonate in a procedure similar to that reported above for dichromate. The reaction is done in fluoride medium, and can be described as:

$$V(OH)_4^+ + 2H^+ + Ag[^{85}Kr] \rightarrow VO^{2+} + Ag^+ + 3H_2O + {}^{85}Kr\uparrow$$

To study the method, a series of solutions was made up as follows: 0.2 ml of NH_4F solution (300 mg/ml) and 0.15 ml of concentrated sulphuric acid were added to 50 ml of ammonium vanadate solutions of various concentrations. A $10^{-7}M$ vanadate solution gave a measurable decrease in the residual activity of the silver kryptonate. It is possible that this sensitivity could be enhanced even more by increasing the time during which the silver kryptonate disk is immersed in the sample [66].

Silver kryptonate may also be used to determine the concentration of *hydrochloric acid*, by a calibration curve technique as in the above methods. However, in this case, it is more convenient to plot residual activity of the silver kryptonate *vs.* the logarithm of the acid concentration. The method might be useful at low concentrations of hydrochloric acid, since an HCl concentration of 10^{-5} M produced a decrease of about 15% in the residual activity of the silver kryptonate [66].

Various types of *analysers* for the determination of *dissolved oxygen* in water (at constant pH) and the determination of *pH* (at constant oxygen concentration) have been developed [59, 62, 63]. The same analysers can be used for the determination of various pollutants with the aid of appropriate radioactive kryptonates. The specific activity of the radioactive kryptonates must be high, since the determinations are based on measurement of the residual activity after contact with the sample.

Analyser for the determination of dissolved oxygen in water, based on discontinuous measurement of the residual activity of thallium kryptonate (*Fig. 6.4*). The lower part of the measuring cell 2 (with the radioactive thallium kryptonate at 3) is counted to determine the initial activity. The time necessary to accumulate 10000 counts is recorded as t_0. With the aid of a suction device 1, the reaction cell 2 is filled with the solution to be analysed (2 ml). After 60 sec, the solution is discharged and the cell disassembled. The dried thallium kryptonate is again counted for the same time t_0. This number of counts can be used to calculate the residual activity as a percentage of the original activity. The oxygen concentration can then be read from an appropriate calibration chart.

Analyser for the determination of dissolved oxygen by measurement of the activity of the water after reaction with the radioactive thallium kryptonate (*Fig. 6.5*). The suction device 1 is used to pull the sample 7 into the reaction cell 4 (2 ml volume), which contains the radioactive thallium kryptonate 5. After a 60-sec contact time, the solution is passed into the measuring cell 10. The activity of the dissolved ^{85}Kr is measured with a scintillation detector, and the oxygen content is determined from the activity by means of a calibration curve. After the measurement, the sample is discharged from the measuring cell through outlet 12.

Analyser for the determination of dissolved oxygen in water by the

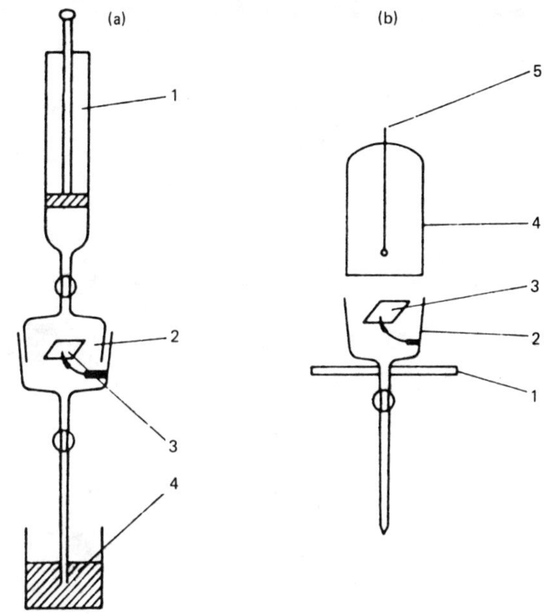

Fig. 6.4 — Analyser for the determination of dissolved oxygen in water, based on discontinuous measurement of the residual activity of the thallium kryptonate:
(a) 1 — suction device (5 ml syringe), 2 — reaction cell (2 ml), 3 — radioactive thallium kryptonate, 4 — analysis solution;
(b) 1 — reaction cell support, 2 — lower part of reaction cell, 3 — radioactive thallium kryptonate, 4 — GM counter, 5 — connection to recording device.

measurement of the activity of ^{85}*Kr released from thallium kryptonate* (*Fig. 6.6*). The suction device 1 is used to pull the sample into the reaction cell 4 (2 ml volume), which contains the thallium kryptonate 5. After 60 sec, the solution is passed to the purging cell 11 (2.5 ml volume), by pressure. Stopcock 17 is closed and stopcock 18 opened. Nitrogen is passed (through inlet 2 and valve 3) to purge the gaseous ^{85}Kr from the aqueous sample into the detection cell 13. After purging, stopcock 18 is closed, and the activity of the gaseous ^{85}Kr in the detection chamber is measured. After the measurement, the aqueous sample is discharged from the purging cell 11 through stopcock 17. The observed activity is used to determine dissolved oxygen by means of a calibration curve.

6.3.2 Radioactive metals

Radioactive metals can be used in radio-release methods for determination of oxidizing agents in aqueous solutions. The decisive factors in the choice of the metal are the following: the metal should not react with water, but should react with oxidizing agents to yield ions which do not form precipitates in aqueous media; the metal should also have a radionuclide with suitable nuclear properties. These conditions are met by thallium (^{204}Tl) and silver (^{110m}Ag), which have been used most frequently for environmental

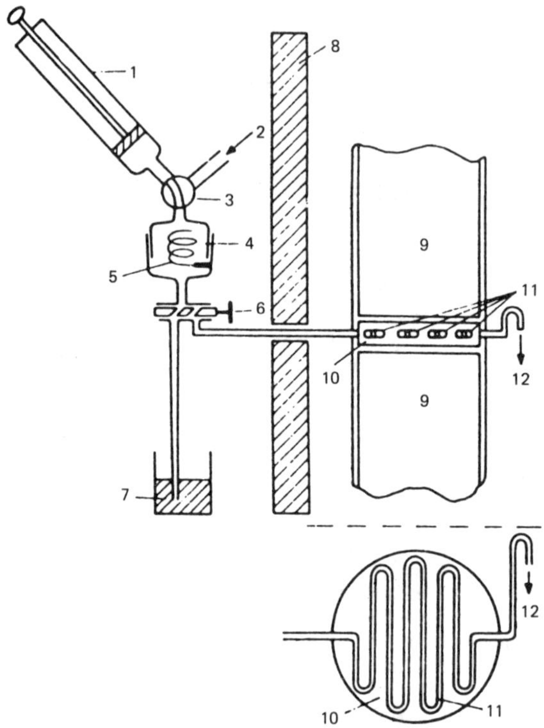

Fig. 6.5 — Analyser for the determination of dissolved oxygen in water by the measurement of ^{85}Kr in the water: 1 — suction device (15 ml syringe), 2 — inlet for nitrogen or air, 3 — three-way stopcock, 4 — reaction cell (2 ml volume), 5 — radioactive kryptonate with homogeneous distribution, 6 — two-way stopcock, 7 — analysis sample, 8 — lead shielding, 9 — photomultiplier, 10 — scintillation measuring cell with a drilled channel, 11 — drilled channel (2 ml), 12 — outlet for analysis solution.

samples. For example, *dissolved oxygen* has been determined in sea-water, drinking water, fresh water and industrial water [67–69]. The determination is based on quantitative oxidation of labelled metallic thallium (^{204}Tl) by dissolved oxygen according to the reaction:

$$4Tl + O_2 + 2H_2O \rightarrow 4Tl^+ + 4OH^-$$

One gram of thallium is needed to analyse approximately 100 air-saturated water samples (a total volume of about 5 litres).

The laboratory equipment for oxygen determination consists of a glass column filled with the labelled metallic thallium and connected to a flow-type GM counter. The column filling is prepared by labelling a solution of thallium nitrate with thallium-204 and electro-depositing labelled thallium from an acetate medium onto copper plates.

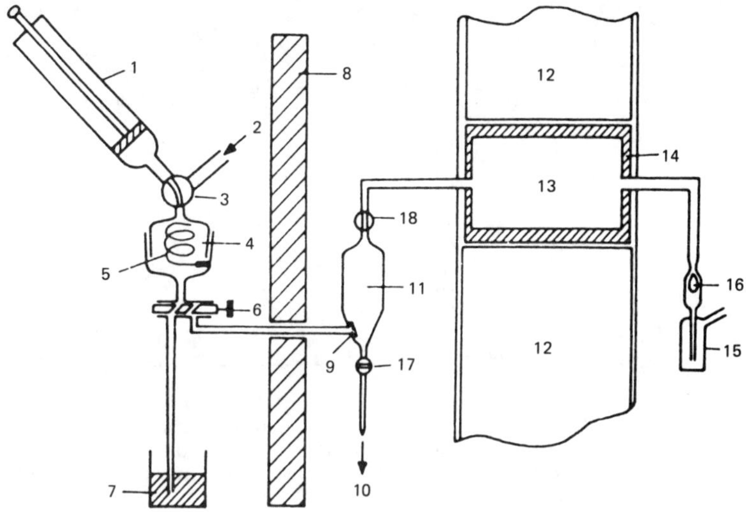

Fig. 6.6 — Analyser for the determination of dissolved oxygen in water by the measurement of the activity of ^{85}Kr released from the aqueous solution. Items 1–8 as in Fig. 6.5, 9 — glass filter, 10 — outlet for analysis solution, 11 — cell for purging of ^{85}Kr, 12 — photomultiplier, 13 — detection cell, 14 — scintillation detector, 15 — water seal, 16 — safety valve, 17, 18 — stopcocks.

The sensitivity of the method is ~0.2 ppm of oxygen in water. The accuracy of the determination is limited by the specific activity involved, and is about ±5% for the 1-ppm level. The calibration graphs for distilled water are linear in the entire range of oxygen concentrations examined and up to about 3 ppm for sea-water. When thallium is deposited on a metal nobler than copper, linearity can also be achieved for oxygen in sea-water.

A portable transistorized oxygen analyser [70, 71] and an analyser for determining oxygen in deep-sea oceanographic research [69] have also been constructed. A block scheme of the analyser for oceanographic studies is shown in Fig. 6.7.

Metallic silver has been used [72] for the determination of *dichromate* on the basis of the reaction:

$$Cr_2O_7^{2-} + 6Ag + 14H^+ \rightarrow 6Ag^+ + 2Cr^{3+} + 7H_2O$$

A piece of radioactive silver metal is used as a stirrer. The reaction takes place in a sulphuric acid medium. Fluoride is used for partial masking of iron, so a Teflon vessel should be used. Before measurement, the solution is neutralized with ammonia. Calibration is done with a standard dichromate solution.

Vanadates have been determined in a column arrangement [73, 74] on the basis of the reaction:

$$V(OH)_4^+ + 2H^+ + Ag \rightarrow VO_2^{2+} + Ag^+ + 3H_2O \ (pH < 3)$$

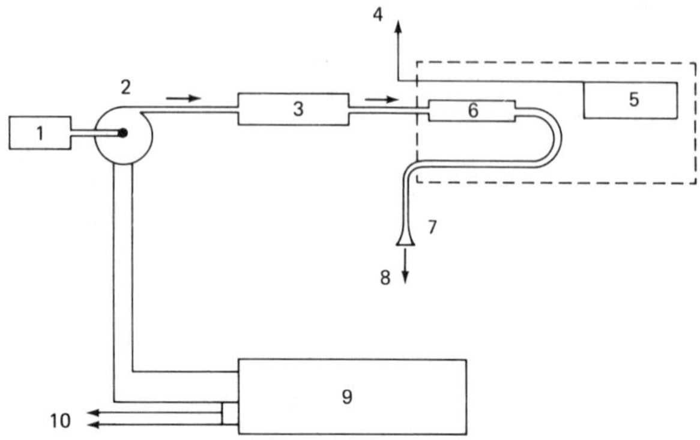

Fig. 6.7 — Diagram of an analyser for oceanographic studies: 1 — input filter, 2 — pump, 3 — thallium column, 4 — coaxial cable to surface, 5 — electronics, 6 — GM counter, 7 — filter, 8 — discharge into sea, 9 — battery and transformer, 10 — control panel.

Vanadium present in a lower oxidation state is oxidized with gaseous chlorine to vanadate. Excess of chlorine is removed by boiling and bubbling nitrogen through the solution. The sensitivity of the determination is 0.1 ppm of vanadium and the reproducibility is $\pm 2\%$ at the 1-ppm level.

6.3.3 Radioactive salts and other radioactive substances

Sulphur dioxide is determined on the basis of the reaction:

$$5SO_2 + 2K^{131}IO_3 + 4H_2O \rightarrow K_2SO_4 + 4H_2SO_4 + {}^{131}I_2$$

which takes place in an alkaline solution through which air containing sulphur dioxide is bubbled. After completion of the reaction, the solution is acidified and the iodine released is extracted [75].

A number of determinations are based on the reaction of a solution of an analyte which can form a soluble complex with a radioactively-labelled precipitate. In this way it is possible to determine anions forming soluble complexes (*e.g.* CN^-, $S_2O_3^{2-}$, I^-, F^-). The principles of the determinations are given in the following equations:

$$2CN^- + {}^{203}Hg(IO_3)_2\downarrow \rightarrow {}^{203}Hg(CN)_2 + 2IO_3^-$$

$$(>0.50\ \mu g/ml,\ \pm 5\%) \quad [76]$$

$$2Cl^- + {}^{203}Hg(IO_3)_2\downarrow \rightarrow {}^{203}HgCl_2 + 2IO_3^-$$

$$(>1\ \mu g/ml,\ \pm 5\%) \quad [77, 78]$$

$$nS_2O_3^{2-} + m\,{}^{110m}AgSCN\downarrow \rightarrow {}^{110m}Ag_m(S_2O_3)_n{}^{(2n-m)-} + mSCN^- \quad [79]$$

$$2I^- + {}^{203}HgI_2\downarrow \rightarrow {}^{203}HgI_4^{2-}$$

$$(>25\ \mu g/ml,\ \pm 5\%) \quad [80]$$

Sometimes the complex is formed with a non-radioactive component of the precipitate [81]:

$$18F^- + 6H^+ + (ZrO)_3({}^{32}PO_4)_2\downarrow \rightarrow 2{}^{32}PO_4^{3-} + 3ZrF_6^{2-} + 3H_2O$$

$$(>4\ \mu g\ F^-,\ \text{paper chromatography})$$

REFERENCES

[1] M. Kyrš, J. Krtil and J. Tölgyessy, *Radioreagent Methods*, in J. Tölgyessy and Š. Varga, *Nuclear Analytical Chemistry*, Vol. 2, *Radioactive Indicators in Chemical Analysis*, Veda, Bratislava; University Park Press, Baltimore 1972.

[2] M. Kyrš, G. N. Bilimovich, J. Tölgyessy *et al.*, *Novye metody radioanaliticheskoi khimii*, Energoizdat, Moscow 1982.

[3] J. Tölgyessy and M. Kyrš, *Radioanalytical Chemistry*, Veda, Bratislava; Ellis Horwood, Chichester, in the press.

[4] M. Kyrš and A. Moravec, *Jaderná energie*, 1984, **30**, 328.

[5] W. J. Driscoll, B. F. Scott and E. A. Huff, *U.S. At. Energy Comm. Rept.*, *TID*-11306, 1961.

[6] A. A. Shatalova, G. I. Meerov, and Ya. R. Savinsky, *Biokhimiya*, 1960, **25**, 577.

[7] B. C. Purkayastha and M. N. Chandra, *J. Indian Chem. Soc.*, 1962, **39**, 231.

[8] N. A. Kiseleva, I. B. Megorskaya and M. I. Rozova, *Zavodsk. Lab.*, 1956, **22**, 1291.

[9] L. E. Mattison and J. C. Wolford, II, *Anal. Chem.*, 1966, **38**, 1675.

[10] J. Starý, *Moderní radioanalytické metody*, in *Nové směry v analytické chemii*, Vol. 1, J. Zýka (ed.), pp. 175–195, SNTL, Prague, 1983.

[11] J. Tölgyessy, Yu. V. Yakovlev and G. N. Bilimovich, *Diagnostika okruzhayushchet sredy radioanaliticheskina metodami*, Energoatomizdat, Moscow, 1985.

[12] T. B. Pierce, *Analyst*, 1960, **85**, 166.

[13] J. Starý, J. Prášilová and T. Bouda, *J. Radioanal. Chem.*, 1975, **28**, 95.

[14] A. Zeman, J. Starý and K. Kratzer, *Radiochem. Radioanal. Lett.*, 1970, **4**, 1.

[15] J. Starý, K. Kratzer and A. Zeman, *J. Radioanal. Chem.*, 1970, **5**, 71.

[16] K. Kratzer, J. Starý, A. Zeman and V. Majer, *Collect. Czech. Chem. Commun.*, 1972, **37**, 3267.

[17] T. B. Handley, *Anal. Chem.*, 1964, **36**, 153.

[18] J. Starý, B. Valter, K. Kratzer and J. Prášilova, *Radiochem. Radioanal. Lett.*, 1977, **30**, 281.

[19] J. Starý and J. Prášilová, *Radiochem. Radioanal. Lett.*, 1976, **26**, 193.

[20] K. Kostadinov and R. Djingova, *Radiochem. Radioanal. Lett.*, 1981, **48**, 401.

[21] J. Starý and J. Prášilová, *Radiochem. Radioanal. Lett.*, 1976, **24**, 143.

[22] H. G. Richter, *Anal. Chem.*, 1967, **38**, 317.

[23] N. Ikeda and S. Amano, *Radioisotopes*, 1967, **16**, 317.

[24] A. Zeman and K. Kratzer, *Radiochem. Radioanal. Lett.*, 1976, **27**, 217.

[25] J. Starý and K. Kratzer, *Int. J. Environ. Anal. Chem.*, 1972, **2**, 79.

[26] K. Kratzer, *Radiochem. Radioanal. Lett.*, 1970, **5**, 69.

[27] J. Starý, K. Kratzer and A. Zeman, *Radiochem. Radioanal. Lett.*, 1971, **6**, 1.

[28] M. Kyrš, *Anal. Chim. Acta*, 1966, **33**, 245.

[29] K. Burger, *Magyar Kém. Folyóirat*, 1964, **70**, 143; *Acta Chim. Acad. Sci. Hung.*, 1964, **40**, 17.

[30] M. H. Kurbatov and J. D. Kurbatov, *J. Am. Chem. Soc.*, 1947, **69**, 438.

[31] M. Kyrš, *Chem. Listy*, 1967, **61**, 440.

[32] T. Braun, J. Tölgyessy and J. Konečný, *Acta Chim. Acad. Sci. Hung.*, 1966, **49**, 131.
[33] A. R. Landgrebe, L. T. McClendon, and J. R. DeVoe, in *Radiochemical Methods of Analysis*, Vol. 2, p. 321, IAEA, Vienna, 1965.
[34] J. Konečný, J. Tölgyessy and T. Braun, *Proc. Anal. Chem. Conf., Budapest*, 1966, **2**, 471.
[35] J. Tölgyessy and Š. Varga, *Talanta*, 1970, **17**, 659.
[36] J. Tölgyessy, Š. Varga and E. H. Klehr, *Radioisotopy*, 1874, **15**, 623.
[37] V. Balek and J. Tölgyessy, *Emanation Thermal Analysis and other Radiometric Emanation Methods*, Elsevier,. Amsterdam, 1984.
[38] J. Tölgyessy and Š. Varga, *Nuclear Analytical Chemistry II, Radioactive Indicators in Chemical Analysis*, University Park Press, Baltimore, Veda, Bratislava, 1972.
[39] J. Tölgyessy and Š. Varga, *Nukleárna analytická chémia*, Alfa, Bratislava, 1976.
[40] Š. Varga and J. Tölgyessy, *Rádiochémia a radiačná chémia*, Alfa, Bratislava, 1976.
[41] J. Tölgyessy, *Radioactive Kryptonates in the Analysis os Environmental Samples*. Consultants' Meeting on Advances in Nuclear Techniques for Analysis of Environmental Samples, Karlsruhe, FRG, 19–21 February 1986.
[42] E. H. Klehr, J. Tölgyessy, M. M. Naoum and J. Pružinec, *Isotopenpraxis*, 1978, **14**, 393.
[43] D. J. Chleck and C. A. Ziegler, in *Radioisotopes in the Physical Sciences and Industry*, p. 351, IAEA, Vienna, 1962.
[44] D. J. Chleck, *U.S. Patent* No. 3 230368, 18 January 1966.
[45] J. Pružinec, J. Tölgyessy and M. M. Naoum, *Radiochem. Radioanal. Lett.*, 1974, **20**, 73.
[46] J. Tölgyessy and J. Pružinec, *Czechoslovak Patent* 166064.
[47] C. O. Hommel, D. J. Chleck and F. J. Brousaides, *Nucleonics*, 1961, **19**, No. 5, 94.
[48] R. L. Bersin, F. J. Brousaides and C. O. Hommel, *J. Air Pollution Control. Assoc.*, 1962, **12**, No. 3, 129, 146.
[49] S. M. Klainer, *The Monitoring of Gaseous Contaminants in the Atmosphere Using Radioactive Clathrate Techniques*, in *Radioisotopes for Aerospace*, Part 2, *Systems and Applications*, *Proc. Symp., Dayton, Ohio 15–17 February* 1966, J. C. Dempsey and P. Polishuk (eds.), pp. 376–94, Plenum Press, New York, 1966.
[50] J. Pružinec and J. Tölgyessy, *Coll. Sci. Papers CHTF*, SVŠT, Bratislava, 1977, 123.
[51] J. Pružinec and J. Tölgyessy, *Jaderná energie*, 1973, **19**, 345.
[52] J. Pružinec and J. Tölgyessy, *Czechoslovak Patent* 160035.
[53] C. O. Hommel, F. J. Brousaides and R. L. Bersin, *Anal. Chem.*, 1962, **34**, 1608.
[54] V. Jesenák, M. M. Naoum and J. Tölgyessy, *Radiochem. Radioanal. Lett.*, 1973, **13**, 199.
[55] V. Jesenák, J. Tölgyessy and M. M. Naoum, *Czechoslovak Patent* 165868.
[56] M. M. Naoum, J. Tölgyessy, V. Jesenák and E. H. Klehr, *J. Radioanal. Chem.*, 1975, **25**, 129.
[57] S. M. Klainer and R. L. Bersin, *Interagency Chemical Rocket Propulsion Group*, *2nd Meeting*, ICRPG Working Group on Hazards, Sacramento, Calif., 9 December 1965.
[58] P. Goodman, *Int. Lab.*, 1971, December, 32.
[59] J. Tölgyessy, E. H. Klehr, P. Dillinger and M. Harangozô, *Isotopenpraxis*, 1980, **16**, 268.
[60] J. Tölgyessy and Š. Varga, *Talanta*, 1969, **16**, 625.
[61] J. Tölgyessy and Š. Varga, *Czechoslovak Patent* 132542.
[62] J. Tölgyessy and P. Dillinger, *Czechoslovak Patent* 166337.
[63] P. Dillinger and J. Tölgyessy, *Czechoslovak Patent* PV 8538–72.
[64] J. Tölgyessy, Š. Varga, V. Jesenák, P. Lukáč and P. Dillinger, *Czechoslovak Patent* 152796.
[65] J. Tölgyessy, Š. Varga, V. Jesenák, P. Lukáč and P. Dillinger, *Radiochem. Radioanal. Lett.*, 1970, **5**, 331.
[66] J. Tölgyessy, V. Jesenák and E. Koval, *J. Radioanal. Chem.*, 1970, **4**, 13.
[67] H. G. Richter and A. S. Gillespie Jr., *Anal. Chem.*, 1962, *Anal. Chem.*, 1116.
[68] J. Ryagert, A. Claus, and C. Beaudet, EUR 2991f, *Euratom Contract* No. 021–64–9, IRAB, Brussels, July 1966.
[69] A. S. Gillespie Jr. and K. F. Roberts, *USAEC Contract* No. AT-(40–1)–2513, Task No. 11, Report No. ORO–596, 31 December 1965.
[70] A. S. Gillespie Jr. and H. G. Richter, *USAEC Contract* No. AT–(40–1)–2513, Task No. 11, Rept. No. ORO–592, 31 December 1963.
[71] A. S. Gillespie Jr. and H. G. Richter, *Analysis Instrumentation* 1965, *Proc. Anal. Instrumentation Div. Symposium*, Instrument Society of America Meeting, Montreal, Canada, May 1965.
[72] H. G. Richter and A. S. Gillespie Jr., *Anal. Chem.*, 1965, **37**, 1146.
[73] A. S. Gillespie Jr. and H. G. Richter, *Anal. Chem.*, 1964, **36**, 2473.
[74] A. S. Gillespie Jr. and H. G. Richter, *Trans. Am. Nuclear Soc.*, 1962, **5**, 273.

[75] H. H. Ross and W. S. Lyon, *Radiochemical Methods of Analysis*, Vol. 2, p. 285, IAEA, Vienna, 1965.
[76] É. Bányai, F. Szabadváry and L. Erdey, *Talanta*, 1963, *Talanta*, 499.
[77] É. Bányai, F. Szabadváry and L. Erdey, *Mikrochim. Acta*, **1962**, 427.
[78] F. Szabadváry, É. Bányai and L. Erdey, *Chim. Anal. Paris*, 1963, **45**, 289.
[79] N. H. Schöön, *Acta Chem. Scand.*, 1960, **14**, 601.
[80] É. Bányai and F. Szabadváry, *Mikrochim. Acta*, **1966**, 601.
[81] J. M. Peixoto-Cabral and H. Götte, *Z. Naturforsch.*, 1955, **10b**, 440.

7

Activation analysis

Activation analysis has become one of the most important and versatile methods of nuclear environmental chemical analysis. This method has opened up new fields of environmental research, reaching new levels of sensitivity for a large number of elements. Most elements of the periodic table can be determined in the μg/g (or μg/ml) range, and some elements at the ng/g (or ng/ml) level. Usually, the highest sensitivities are obtained by using a nuclear reactor, since fluxes up to 10^{15} n . cm^{-2} . sec^{-1} are available. Elements which cannot be determined by neutron irradiation can sometimes be determined by applying another type of activation, for instance, by photons or charged particles (alpha particles, deuterons, protons, etc.). Many books dealing with activation analysis are available [1–13].

7.1 BASIC PRINCIPLES OF ACTIVATION ANALYSIS

Activation analysis is a method of elemental analysis based on the measurement of characteristic radiation from radionuclides formed directly or indirectly by activation. In general, a specification is added as to the type of the incident radiation to be used (neutron, photon, charged particles).

Activation analysis requires a consideration of the laws governing growth and decay during activation. The final basic equation is:

$$A = \frac{m}{M} N_A \phi \sigma \theta S D C P_\gamma \varepsilon \qquad (7.1)$$

where A = measured activity (Bq),
m = amount of element determined (g),
M = atomic weight of the element to be determined (g/mole),
N_A = Avogadro constant,
ϕ = activating flux (particles . m^{-2} . sec^{-1}),

σ = activation cross-section of reaction (m^2),
θ = relative natural isotopic abundance of the activated isotope,
S = $1 - \exp(-\lambda t_1)$ = saturation factor (λ = decay constant of radioactive product, t_1 = irradiation time, sec),
D = $\exp(-\lambda t_2)$ = decay factor, for correction of final activity after cooling time t_2 (from the end of activation to the beginning of measurement of activity),
C = $(1 - \exp[-\lambda t_3]/\lambda t_3)$ = factor for correction of decay of nuclide during measuring time t_3,
P_γ = probability of emission of a photon with energy E_γ,
ε = detection efficiency for the measured radiation energy.

From this equation it is obvious that the measured activity A, and thus the sensitivity of the determination, is directly proportional to the flux of activating particles, ϕ, and to the cross-section of the activation reaction used, σ. Thus analysts attempt to use fluxes that are as high as possible; therefore reactor neutron-activation analysis is the most important technique. The cross-section is a function of particle energy and thus it is necessary to select an energy that will give a high activation cross-section value.

The sensitivity can be raised by prolonging the activation time and thus approach saturation activity. However, this prolongation is only of limited significance.

In addition to the factors mentioned, the properties of the radiation emitted by the radionuclide formed, and the detection efficiency, also affect the sensitivity.

Activation analysis may be subdivided, according to the steps involved, into two different classes:

(1) non-destructive activation analysis,
(2) destructive (radiochemical) activation analysis.

In non-destructive activation analysis (NDAA), after the irradiation, no chemical and physical operations are applied which cause a change of any properties of the sample.

When the radionuclides produced cannot be differentiated instrumentally, a chemical separation may be needed (radiochemical activation analysis, RAA).

A special and often-used method of activation analysis is instrumental activation analysis (IAA) in which element specificity is obtained by using appropriate irradiation conditions, radiation measurement techniques and mathematical techniques for the interpretation of the results. The term IAA is often used as a synonym for non-destructive activation analysis.

Figure 7.1 illustrates the steps in activation analysis [14]. IAA is applied whenever possible, because it involves fewest operations: sampling, irradiation and radioactivity measurement. By gamma-spectrometry a great number of radionuclides can be measured simultaneously. Thus, IAA in

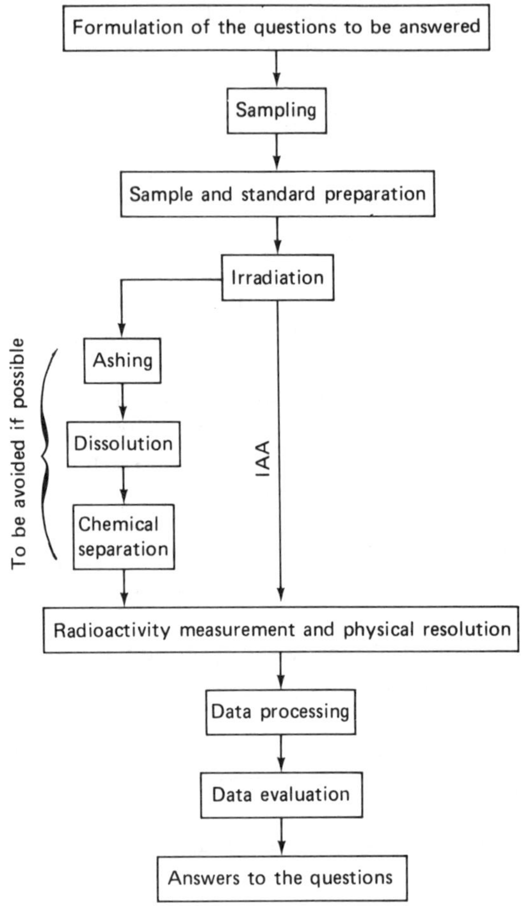

Fig. 7.1 — General scheme of activation analysis.

combination with gamma-spectrometry is an excellent multielement method. If gamma-lines of radionuclides cannot be separated in the spectrum, or if the radionuclides cannot be separated in the spectrum, or if the radionuclides that are to be measured do not emit gamma-rays, radiochemical separation methods have to be applied. In this case, an important advantage of activation analysis over other methods of trace element analysis is the fact that impurities introduced into the system after irradiation do not influence the results, because they are not activated [15].

Activation analysis involves the following steps [16].

(1) Investigation of the feasibility of the technique, taking into account:
- — type of activation, sensitivity of technique and interfering reactions,
- — availability of activating source and radiation measuring devices,
- — cost of activation technique *vs*. cost by non-nuclear techniques.

(2) Sample preparation:
- — suitable containers are chosen and loaded with aliquots of samples and comparator or standards (aluminium, silica, polyethylene and polystyrene containers are commonly used).

(3) — Irradiation:
- — samples and comparator or standards are irradiated with as nearly the same flux as possible.

(4) — Purification:
- — if analysis by direct gamma-spectrometry is possible, the materials are transferred to non-irradiated counting containers,
- — if necessary, the materials are dissolved, inactive carriers are added and purification done by radiochemical techniques.

(5) — Comparative measurement:
- — intensities of radiation from samples and standards are determined, corrected for decay, chemical yield, counting efficiency, self-shielding during irradiation, and correlated with weight of element in the samples.

7.1.1 Qualitative activation analysis

After activation of the sample, the radionuclides are identified by physical methods on the basis of the type, half-life and energy of the radiation emitted. However, in more complicated mixtures difficulties are encountered in a systematic identification and the physical methods should be combined with chemical separations for identification of radionuclides.

Gamma-spectrometry is the method most often used for the identification of radionuclides. A reliable identification of elements from gamma-spectra of radionuclides formed by neutron activation based on (n,γ)-reactions is possible, if the following data are available:

- — energy of gamma lines,
- — number and relative intensities of gamma lines,
- — decay of the intensity of gamma lines with time,
- — gamma-spectra of several radionuclides of the given element if the element has several stable isotopes.

In addition to this, the identification can be performed by using other nuclear reactions, such as (n,p), $(n,2n)$ and (n,α).

Investigation of the gamma-spectrum obtained according to the first three data items above is usually sufficient for reliable identification. Radionuclide identification can be facilitated by using specialized tables and atlases of standard gamma-spectra obtained with scintillation as well as semiconductor Ge(Li) gamma-spectrometers.

Pelekis [17] has provided a table (Table 7.1) of radioactive isotopes obtained by (n,γ) reactions, where isotopes are classified according to gamma-line energies and half-lives. The gamma-lines taken into account were those with relative intensity that was not $<0.1\%$ of that of the most

Table 7.1 — Characteristics of certain radionuclides obtained by (n, γ)-reactions

t \ E MeV	0.02–0.04	0.04–0.06	0.06–0.08	0.08–0.1	0.1–0.2	0.2–0.4
0.1–1 sec	^{181m}Ta	^{73m}Ge				
1–10 sec			^{183m}W	^{178m}Hf	^{183m}W, ^{114m_2}In, ^{116m_2}In, ^{177m}Yb	^{167m}Er, ^{177m}Yb, ^{178m}Hf
10 sec–1 min	^{169m}Yb, ^{199m}Pt		^{125m}Xe	^{109m}Ag	^{19}O, ^{46m}Sc, ^{75m}Ge, ^{77m}Se, ^{125m}Xe, ^{194m}Ir, ^{81m}Kr	^{179m}Hf, ^{194m}If, ^{199m}Pt, ^{107m}Pd
1–10 min	^{124m}Sb	^{94m}Nb, ^{161}Gd, ^{190m}Os, ^{129m}Ir	^{122m}Sb, ^{161}Gd, ^{185m}W	^{79m}Se	^{27}Mg, ^{71}Zn, ^{109m}Pd, ^{127m}Xe, ^{161}Gd, ^{165m}Dy, ^{185m}W, ^{190m}Os	^{51}Ti, ^{125m}Sn, ^{161}Gd, ^{165m}Dy, ^{137}Xe, ^{190m}Os, ^{205}Hg
10 min–1 hr		^{60m}Co	^{188m}Re, ^{199}Pt, ^{239}U	^{101}Mo, ^{151}Nd, ^{233}Th	^{70}Ga, ^{81m}Sc, ^{83}Sc, ^{101}Mo, ^{111m_2}Cd, ^{101}Tc*, ^{116m_1}In, ^{123}Sn, ^{131}Te, ^{151}Nd, ^{155}Sm, ^{182m}Ta, ^{188m}Re, ^{199}Pt, ^{199m}Hg, ^{233}Th	^{83}Se, ^{101}Tc*, ^{111}Pd, ^{151}Nd, ^{155}Sm, ^{182m}Ta, ^{199}Pt, ^{199m}Hg, ^{233}Th
1–10 hr	^{80m}Br, ^{83}Br*, ^{129}Te, ^{134m}Cs, ^{149}Nd, ^{189m}Os	^{80m}Br, ^{83}Br*, ^{127}Te, ^{180m}Hf	^{75}Ge, ^{157}Dy	^{149}Nd, ^{157}Dy, ^{165}Dy, ^{171}Er, ^{176m}Lu, ^{180m}Hf	^{75}Ge, ^{85m}Kr, ^{105}Ru, ^{111m}Pd, ^{117}In*, ^{134m}Cs, ^{139}Ba, ^{149}Nd, ^{152m}Eu, ^{157}Dy, ^{171}Er, ^{177}Yb	^{65}Ni, ^{71m}Zn, ^{75}Ge, ^{85m}Sr, ^{85m}Kr, ^{87m}Sr, ^{93m}Mo, ^{105}Ru, ^{107}Cd*, ^{117m}Cd, ^{127}Te, ^{135}Xe, ^{149}Nd, ^{152m}Eu, ^{157}Dy, ^{165}Dy, ^{171}Er, ^{180m}Hf, ^{197m}Pt
10 hr–1 days		^{125}Xe, ^{159}Gd	^{187}W, ^{191m}Os, ^{197}Pt	^{109}Pd, ^{125}Xe, ^{159}Gd	^{125}Xe, ^{187}W, ^{188}Re, ^{197}Pt, ^{197m}Hg	^{42}K, ^{77}Ge, ^{109}Pd, ^{125}Xe, ^{159}Gd, ^{194}Ir, ^{197}Pt
1–10 days	^{161}Tb*	^{79}Kr, ^{99}Mo, ^{129m}Xe, ^{143}Ce, ^{161}Tb*, ^{199}Au*	^{153}Sm, ^{161}Tb*, ^{177}Lu, ^{193}Os, ^{197}Hg	^{77}As, ^{79}Kr, ^{131m}Te, ^{133}Xe, ^{161}Tb*, ^{166}Dy, ^{191}Pt, ^{195m}Pt	^{47}Sc*, ^{79}Kr, ^{97}Ru, ^{99}Mo, ^{129m}Xe, ^{131m}Te, ^{143}Ce, ^{153}Sm, ^{151}Pm*, ^{161}Tb*, ^{175}Yb, ^{177}Lu, ^{186}Re, ^{191}Pt, ^{193}Pt, ^{195m}Pt, ^{197}Hg, ^{199}Au*, ^{77}As*	^{77}As*, ^{79}Kr, ^{97}Ru, ^{99}Mo, ^{105}Rh*, ^{115}Cd, ^{125}Sn, ^{131m}Te, ^{131}I*, ^{133m}Xe, ^{140}La, ^{137m}Ce, ^{143}Ce, ^{149}Pm*, ^{151}Pm*, ^{133m}Ba, ^{135m}Ba, ^{175}Yb, ^{177}Lu, ^{191}Pt, ^{193}Os, ^{199}Au*
10–100 days	^{125}I*, ^{125m}Te, ^{169}Yb	^{103}Ru, ^{103}Pd, ^{127}Xe, ^{191}Os	^{103}Pd, ^{121}Te, ^{169}Yb	^{131}Ba, ^{147}Nd, ^{160}Tb, ^{175}Hf, ^{193m}Ir	^{59}Fe, ^{85}Sr, ^{114m}In, ^{117m}Sn, ^{125m}Te, ^{127}Xe, ^{131m}Xe, ^{131}Ba, ^{141}Ce, ^{147}Nd, ^{160}Tb, ^{169}Yb, ^{175}Hf, ^{181m}Hf, ^{185}Os, ^{191}Os	^{51}Cr, ^{95}Zr, ^{103}Pd, ^{111}Ag*, ^{117m}Sn, ^{127}Xe, ^{131}Ba, ^{147}Nd, ^{160}Tb, ^{169}Y, ^{175}Hf, ^{185}Os, ^{192}Ir, ^{203}Hg
100 days–1 yr		^{127m}Te, ^{159}Dy	^{75}Se, ^{145}Sm, ^{153}Gd, ^{170}Tm	^{75}Se, ^{145}Sm, ^{153}Gd, ^{170}Tm, ^{182}Ta	^{75}Se, ^{123m}Te, ^{139}Ce, ^{153}Gd, ^{181}W, ^{182}Ta	^{75}Se, ^{113}Sn, ^{121m}Te, ^{139}Ce, ^{182}Ta
1–10 yr	^{155}Eu*	^{133}Ba, ^{155}Eu*	^{155}Eu*, ^{171}Tm*	^{108}Ag, ^{133}Ba, ^{155}Eu*	^{125}Sb*	^{125}Sb*, ^{133}Ba
> 10 yr	^{151}Sm		^{166m}Ho		^{85}Kr, ^{152}Eu, ^{154}Eu, ^{166m}Ho	^{155}Eu, ^{154}Eu, ^{166m} Ho

Table 7.1 — *Contd.*

t \ E MeV	0.4–0.6	0.6–0.8	0.8–1.0	1.0–1.2	1.2–1.4	1.4–1.8	1.8–2.5	>2.5
0.1–1 sec	^{175m}Yb, ^{207m}Pb	^{38m}Cl, ^{135m2}Ba		^{207m}Pb				
1–10 sec	^{178}Hf						^{16}N	^{15}C, ^{16}N
10 sec–1 min	^{23}Ne, ^{100}Tc, ^{104}Rh, ^{197m}Au	^{100}Tc, ^{110}Ag, ^{139m}Ce, ^{194m}Ir	^{100}Tc, ^{110}Ag	^{100}Tc	^{19}O, ^{100}Tc, ^{104}Rh	^{20}F, ^{23}Ne, ^{100}Tc	^{23}Ne	
1–10 min	^{71}Zn, ^{86m}Rb, ^{104m}Rh, ^{108}Ag, ^{114}In, ^{137}Xe, ^{161}Gd	^{51}Ti, ^{104m}Rh, ^{108}Ag, ^{114}In, ^{125}Sn, ^{190m}Os, ^{137m}Ba	^{27}Mg, ^{51}Ti, ^{71}Zn, ^{104m}Rh	^{27}Mg, ^{66}Cu, ^{71}Zn, ^{125m}Sn	^{104m}Rh, ^{114}In, ^{125m}Sn	^{28}Al, ^{52}V, ^{104m}Rh		^{37}S, ^{49}Ca
10 min–1 hr	^{101}Mo, ^{101}Tc*, ^{111}Pd, ^{116m}In, ^{117}Cd, ^{131}Te, ^{135m}Xe, ^{151}Nd, ^{128}I, ^{199}Pt, ^{233}Th	^{80}Br, ^{88}Rb, ^{101}Mo, ^{101}Tc*, ^{111}Pd, ^{128}I, ^{131}Te, ^{151}Nd, ^{199}Pt, ^{233}Th	^{88}Rb, ^{101}Mo, ^{101}Tc*, ^{116m}In, ^{111}Pd, ^{128}I, ^{131}Te, ^{199}Pt, ^{233}Th	^{70}Ga, ^{101}Mo, ^{116m}In, ^{131}Te, ^{151}Nd	^{70}Ga, ^{83}Rb, ^{111}Pd, ^{116m}In, ^{151}Nd	^{38}Cl, ^{101}Mo, ^{111}Pd, ^{116m}In, ^{151}Nd	^{38}Cl, ^{88}Rb, ^{101}Mo, ^{116m}In, ^{151}Nd	^{88}Rb
1–10 hr	^{71m}Zn, ^{75}Ge, ^{87}Kr, ^{105}Ru, ^{107}Cd*, ^{117m}Cd, ^{117}In*, ^{127}Te, ^{129}Te, ^{137}Ce, ^{149}Nd, ^{163}Er, ^{165}Dy, ^{180m}Hf	^{71m}Zn, ^{75}Ge, ^{93m}Mo, ^{97}Nb*, ^{105}Ru, ^{107}Cd*, ^{135}Xe, ^{149}Nd, ^{165}Dy	^{56}Mn, ^{87}Kr, ^{107}Cd*, ^{117m}Cd, ^{105}Ru, ^{152m}Eu, ^{165}Dy, ^{177}Yb	^{65}Ni, ^{97}Nb*, ^{107}Cd*, ^{129}Te, ^{163}Er, ^{165}Dy, ^{177}Yb	^{31}Si, ^{41}As, ^{105}Ru, ^{107}Cd, ^{117m}Cd, ^{139}Ba, ^{152m}Eu, ^{177}Yb	^{65}Ni, ^{93m}Mo, ^{105}Ru, ^{111m}Pd, ^{117m}Cd, ^{139}Ba	^{56}Mn, ^{87}Kr, ^{117m}Cd	^{56}Mn, ^{87}Kr
10 hr–1 days	^{64}Cu*, ^{69m}Zn, ^{77}Ge, ^{125}Xe, ^{130}I, ^{187}W, ^{188}Re	^{72}Ga, ^{77}Ge, ^{97}Zr, ^{130}I, ^{187}W, ^{188}Re, ^{194}Ir	^{72}Ga, ^{77}Ge, ^{187}W, ^{183}Re, ^{194}Ir	^{72}Ga, ^{77}Ge, ^{130}I, ^{183}Re, ^{194}Ir	^{24}Na, ^{64}Cu*, ^{77}Ge, ^{97}Zr, ^{188}Re, ^{194}Ir	^{42}K, ^{72}Ga, ^{77}Ge, ^{97}Zr, ^{142}Pr, ^{188}Re, ^{194}Ir	^{72}Ga, ^{77}Ge, ^{97}Zr, ^{188}Re, ^{194}Ir	^{24}Na, ^{72}Ga, ^{97}Zr
1–10 days	^{47}Ca, ^{76}As, ^{77}As*, ^{79}Kr, ^{82}Br, ^{97}Ru, ^{115}Cd, ^{122}Sb, ^{125}Sn, ^{131m}Te, ^{140}La, ^{143}Ce, ^{151}Pm*, ^{153}Sm, ^{166}Dy, ^{191}Pt, ^{193}Os, ^{198}Au, ^{183}Ta, ^{105}Rh*	^{76}As, ^{82}Br, ^{79}Kr, ^{99}Mo, ^{122}Sb, ^{131}I*, ^{131m}Te, ^{143}Ce, ^{151}Pm*, ^{153}Sm, ^{186}Re, ^{191}Pt, ^{198}Au	^{47}Ca, ^{82}Br, ^{125}Sn, ^{131m}Te, ^{140}La, ^{143}Ce	^{82}Br, ^{122}Sb, ^{125}Sn, ^{131m}Te, ^{143}Ce, ^{198}Au	^{47}Ca, ^{76}As, ^{82}Br, ^{131m}Te, ^{122}Sb, ^{166}Ho	^{76}As, ^{82}Br, ^{125}Sn, ^{131m}Te, ^{140}La, ^{166}Ho	^{76}As, ^{125}Sn, ^{131m}Te, ^{140}La	
10–100 days	^{85}Sr, ^{103}Ru, ^{103}Pd, ^{114m}In, ^{115m}Cd, ^{121}Te, ^{131}Ba, ^{147}Nd, ^{175}Hf, ^{181}Hf, ^{185}Os, ^{192}Ir	^{95}Zr, ^{95}Nb*, ^{103}Ru, ^{124}Sb, ^{114m}In, ^{131}Ba, ^{147}Nd, ^{160}Tb, ^{185}Os, ^{192}Ir	^{46}Sc, ^{115m}Cd, ^{124}Sb, ^{131}Ba, ^{160}Tb, ^{185}Os	^{46}Sc, ^{115m}Cd, ^{124}Sb, ^{131}Ba, ^{160}Tb, ^{185}Os	^{59}Fe, ^{115m}Cd, ^{124}Sb, ^{160}Tb	^{86}Rb, ^{124}Sb	^{124}Sb	
100 days–1 yr	^{65}Zn*, ^{75}Se, ^{110m}Ag, ^{145}Sm	^{110m}Ag, ^{127m}Te	^{110m}Ag, ^{182}Ta	^{65}Zn*, ^{121m}Te, ^{123m}Sn, ^{182}Ta	^{182}Ta	^{110m}Ag, ^{182}Ta	^{110m}Ag	
1–10 yr	^{108m}Ag, ^{125}Sb*, ^{134}Cs	^{108m}Ag, ^{125}Sb*, ^{134}Cs		^{60}Co, ^{134}Cs	^{60}Co, ^{134}Cs			
> 10 yr	^{85}Kr, ^{152}Eu	^{94}Nb, ^{152}En	^{94}Nb, ^{152}Eu, ^{154}Eu, ^{166m}Ho	^{152}Eu, ^{154}Eu	^{154}Eu	^{94}Nb, ^{152}Eu, ^{154}Eu		

intense gamma-line (insofar as the intensities of the gamma-lines were known). Nuclides obtained through the radioactive decay of short-lived radionuclides, and those showing an intense positron decay (^{64}Cu, ^{65}Zn, ^{79}Kr, ^{109}Cd) are marked with an asterisk.

7.1.2 Quantitative activation analysis

7.1.2.1 Relative method

The radiochemical purity of standards and samples is established by counting under identical conditions, by beta-counting, gamma-spectrometry, alpha-counting, or other suitable measuring techniques. Half-lives of nuclides, absorption characteristics and gamma-spectra are aids in establishing the identity of the radionuclides. Finally, the activities of the standards and samples are compared and the mass of unknown element (X) in the sample is calculated by means of the relation

$$\frac{\text{mass of X in sample}}{\text{mass of X in standard}} = \frac{\text{activity of X in sample}}{\text{activity of X in standard}} \tag{7.2}$$

The activities used in the equation must have been corrected for decay and other factors.

In the case of multielement analysis this classical procedure has some obvious disadvantages, which can be summarized as follows [18].

(1) The preparation, irradiation and measurement of a large number of standards (usually only for expected elements) is time-consuming and may introduce errors.
(2) Quantitative analysis is impossible for unexpected elements because no standards are provided.
(3) The calculation of detection limits is impossible, owing to lack of sufficient standards.
(4) It is sometimes difficult to ensure identical irradiation conditions for standards and samples; problems may arise from flux inhomogeneities along the irradiation can, variation of particle flux during repeated irradiations, self-shielding effects in the standards and lack of space in the irradiation capsule.
(5) Handling of a large number of standards causes great difficulties in computer-coupled, automated activation analysis.

The difficulties mentioned above can be partially eliminated by using synthetic multielement standards or standard reference materials [19–21].

Single-element standards are easier to prepare and handle, but the ratios of the effective cross-sections (ratios of the resonance integrals) of the standard and the elements to be determined must be well known. Multielement standards must contain known amounts of all elements to be determined and must be prepared with care. The availability of homogeneous and reliable multielement standards is of great importance in activation analysis. All standards must be irradiated under as nearly as possible the same conditions as the samples, a condition that can often be more easily fulfilled with small single-element standards [15].

7.1.2.2 Semi-relative (comparator) method

Multielement analysis can be performed by irradiating and measuring a single element (comparator) instead of standards prepared from known weights of the elements to be determined [18]. The specific activities of the elements determined, *a*, can be calculated by multiplying the measured specific activity of the comparator a^* by the experimentally determined *k* factors, which can be written as [21]:

$$k = \frac{a}{a^*} = \frac{M^* P_\gamma \varepsilon \theta \phi a}{M P_\gamma^* \varepsilon^* \theta^* \phi^* \sigma^*} \tag{7.3}$$

where $a = A/SDCm$. The symbols in Eq. (7.3) have the same meaning as those in Eq. (7.1).

In practice the comparator is usually a small thin foil or a few millimetres of wire. Owing to its small dimensions, a comparator can be placed very close to the sample during irradiation, thereby ensuring that $\phi/\phi^* = 1$.

Although very precise and laborious experimental work is needed to determine the *k* factors, once they are available the preparation of standards can be omitted in any further analysis. It is assumed, however, that the *k* values are constant, i.e. that there are no variations in the quantities given in Eq. (7.3) between the time of determination of the *k* values and that of the actual analysis. The accuracy of the method, therefore, depends mainly on the constancy of the *k* values.

This method, based on experimentally-determined and convertible comparison factors (*k*), is considered to be the most suitable for general use. Simonits and co-workers [18,22] proposed an alternative method by introducing generalized k_0 factors which are independent of irradiation and measuring conditions. This approach combines the simplicity of the absolute methods with nearly the same accuracy as the relative ones. It is suggested that k_0 factors be compiled in all cases when using single-comparator methods, to allow a continuous re-evaluation.

7.1.2.3 Absolute method

It can be seen from Eq. (7.1) that when the values of M, θ, λ, P_γ, ϕ and σ are known, the absolute amount of the element present in the sample can be calculated on the basis of the measured value of the absolute activity. Difficulty in determining the flux of bombarding particles occurs in absolute activation analaysis, however, particularly when charged particles from accelerators are used (protons, deuterons, alpha-particles).

The absolute method is very complicated. The determination always has lower accuracy than the relative and semi-relative methods commonly used, where all unreliable factors associated with the flux of activating particles, activation time, activation cross-section and detection efficiency are avoided.

7.1.3 Activation

Activation parameters can be changed over wide ranges, which allows for the selection of the optimum conditions for analysis, e.g. the type of particle, intensity, energy and activation time can be chosen.

All types of particles can be used for activation. However, electrically charged particles have found only restricted applications, owing to the low cross-sections. By contrast, neutron activation analysis has found the widest application because of the high cross-sections of reactions with neutrons. The (n,γ)-reaction is the most important in activation analysis, but applications of (n,α), (n,p) and $(n,2n)$ reactions are also known. Activation with gamma-rays can also be used in certain cases, through the nuclear photo-effect.

The type of activation reaction has to be selected in each case. Ideally the reaction chosen would result in activation of only the element of interest. Chemical separations would then be unnecessary, and the whole analytical procedure simplified to activation, and measurement of the induced activity. In general, the following requirements are considered when selecting the activation reaction.

(1) A suitable activation product has to be formed:
   — its half-life should be sufficiently long to allow for the necessary treatment and/or transport of samples after activation. However, this half-life should not be too long, so that the activation time need not be too prolonged,
   — decay parameters should correspond to the measuring equipment intended to be used.
(2) The reaction has to exhibit optimum sensitivity.
(3) Interfering reactions forming or consuming the reaction products must not occur.
(4) Treatment of the irradiated material must be as simple as possible. Chemical separation can often be avoided by a proper combination of individual factors.

7.2 BASIC TYPES OF ACTIVATION ANALYSIS

7.2.1 Thermal neutron activation analysis

Thermal neutron activation analysis is the most important technique, because high neutron fluxes are available in nuclear reactors and the majority of elements can be activated by thermal neutrons.

In Table 7.2 a survey of the detection limits obtained by irradiation in a nuclear reactor is presented [15]. From this table it is evident that several elements are not activated in measurable amounts by thermal neutrons (e.g. H, C, N), some give only very short-lived radionuclides (e.g. O, F) and from others only beta-emitters are produced (e.g. Si, P, S, Ca, Pb, Bi). Therefore, these elements are not detected by gamma-spectrometry. A great number of the elements mentioned (H, C, N, O, Si, P, S) are the main constituents of water, aerosols, soils and biological materials, and this is the reason why instrumental neutron activation analysis (INAA) is very suitable for the determination of trace elements in environmental samples. However, several elements significant in environmental chemical analysis are not detected by INAA because only beta-emitters are produced (Pb, Bi) and for others the detection limits are relatively high (Cd, Hg).

Important parameters in activation analysis are the irradiation time and the decay time between the end of irradiation and the measurement (see Table 7.2). By variation of the irradiation time, optimal ratios of activities can be obtained. In a similar way, the activity ratios of analytes to interfering elements vary with decay time, and optimal conditions can be selected.

By using three different irradiation times, about 40 elements can be determined.

(1) Short irradiation (several minutes) and short decay time (several minutes): Al, Ca, Cl, Cu, Dy, I, In, Mg, Mn, S, Sr, Ti, V.
(2) Medium irradiation (about 1 hr) and medium decay time (about 1 day): As, Au, Br, Cu, Eu, Ga, K, La, Na, Sb, Sm, W.
(3) Long irradiation (several days) and long decay time (from one to several weeks): Ag, Ba, Ce, Co, Cr, Cs, Eu, Fe, Hf, Hg, Lu, Ni, Rb, Sb, Sc, Se, Ta, Tb, Th, Yb, Zn.

More elements can be determined after radiochemical separation (radiochemical activation analysis) [15].

7.2.2 Fast neutron activation analysis

For elements which cannot be determined by thermal neutron activation with sufficient sensitivity (e.g. O, N, S), activation with fast neutrons (14 MeV) produced in a neutron generator can be used.

Cross-sections of (n, p) and (n, α) reactions with 14 MeV neutrons are higher for light atoms. Those of (n, 2n) reactions increase with increasing atomic number so their values for elements with atomic numbers above 50 are commonly about 1–2 barn. The (n, γ) reactions with fast neutrons show negligible cross-section values and are quite insignificant for analytical purposes.

In Table 7.3 radionuclides are listed according to the sensitivity of their determination [23].

Table 7.2 — Detection limits for determination of elements by neutron activation analysis at a flux density of 10^{14} n.cm^{-2}.sec^{-1}.
(It is assumed that 10 Bq is sufficient to allow quantitative determination). (Reproduced from [15] by permission of Professor K. H. Lieser).

Detection limit (g)	Irradiation time 1 hr	Irradiation time 1 week
10^{-14}–10^{-13}	Dy^{1}	Eu^{1}, Dy^{1}
10^{-13}–10^{-12}	Co, $Rh^{*,2}$, $Ag^{*,2}$, In^{1}, Eu^{1}, Ir	Mn, Co, $Rh^{*,1}$, $Ag^{*,2}$, In, Sm^{1}, Ho, Re^{1}, Ir, Au
10^{-12}–10^{-11}	V, Mn, Se*, Br^{1}, I^{1}, Pr, Er* Yb*, Hf*, Th^{1}	Na, Sc, V, Cu^{2}, Ga^{1}, As, Se*, Br^{1}, Pd Sb, I^{1}, Cs, La, Pr, Er*, Tm^{1}, Yb*, Lu, Hf*, W, Hg, Th^{1}
10^{-11}–10^{-10}	Mg, Al, Cl^{1}, Ar, Cu^{1}, Ga^{2}, Nb, Cs, Sm, Ho, Lu, Re, Au, U	Mg, Al, Cl^{1}, Ar, K^{1}, Cr^{1}, Ni^{1}, Ge^{1}, Kr, Y^{β}, Nb, Ru, Gd^{1}, Tb^{1}, Tl^{1}, Os^{1}, U
10^{-10}–10^{-9}	F*, Na, Ge^{1}, As, Kr, Rb^{1}, Sr, Mo, Ru, Pd, Sb, Te^{1}, Ba, La, Nd^{1}, Gd^{1}, W, Os, Hg, Tl^{β}	F*, P^{β}, Zn, Rb^{1}, Sr, Mo, Te^{1}, Ba, Ce, Nd, Pt, Tl^{β}
10^{-9}–10^{-8}	Ne*, Si^{β}, K, Sc, Ti, Ni, Y^{β}, Cd, Sn, Xe, Tb^{1}, Tm, Ta, Pt	Ne*, Si^{β}, Ti, Cd, Sn, Xe, Bi^{β}
10^{-8}–10^{-7}	P^{β}, Cr^{1}, Zn, Ce	S^{β}, Ca^{β}, Fe, Zr
10^{-7}–10^{-6}	S^{β}, Zr, Pb^{β}, Bi^{β}	Pb^{β}
10^{-6}–10^{-5}	O*, Ca^{β}	O*

*The radionuclides produced from these elements have half-lives ranging from 1 sec to 1 min. Therefore, for quantitative determination, a decay rate of the order of 100 Bq is necessary and the elements have been placed in the next group.

$^{\beta}$Only beta-rays, no gamma-rays.

^{1,2}The percentages of the gamma-transitions are low. Therefore, if the activity is measured with a gamma-spectrometer, the detection limits are lower by the number of orders of magnitude indicated by the superscript.

7.2.3 Activation with charged particles

The element to be determined can be activated not only with neutrons but also with charged particles (protons, deuterons, alpha-particles, tritons, and light helium nuclei). Radionuclides are thus formed through reactions which are different from those obtained with neutrons, and some elements that undergo no activation with neutrons can be determined by activation with charged particles.

Charged-particle activation analysis has been used for many years for the determination of light elements. Recently, various techniques based on the measurement of prompt radiation have enabled structural information, depth distributions and small-volume analyses to be obtained, in addition to

Table 7.3 — Sensitivity of fast neutron (14 MeV) activation analysis ($\phi = 10^9$ n.cm^{-2}.sec^{-1})

Sensitivity (g)	n,p	n,α	n,2n	n,γ
10^{-7}–10^{-6}	^{19}F, ^{28}Si	^{31}P	^{63}Cu, ^{71}Ga, ^{79}Br, ^{109}Ag, ^{121}Sb, ^{141}Pr	
10^{-6}–10^{-5}	^{16}O, ^{23}Na, ^{27}Al, ^{51}V, ^{52}Cr, ^{56}Fe, ^{66}Zn, ^{70}Ge, ^{97}Mo	^{11}B, ^{37}Cl, ^{55}Mn, ^{69}Ga, ^{103}Rh	^{14}N, ^{19}F, ^{39}K, ^{64}Zn, ^{69}Ga, ^{76}Ge, ^{81}Br, ^{82}Se, ^{90}Zr, ^{92}Mo, ^{107}Ag, ^{130}Te, ^{181}Ta, ^{198}Pt	
10^{-5}–10^{-4}	^{24}Mg, ^{25}Mg, ^{29}Si, ^{31}P, ^{34}S, ^{37}Cl, ^{41}K, ^{64}Zn, ^{75}As, ^{88}Sr, ^{105}Pd	^{27}Al, ^{30}Si, ^{34}S, ^{41}K, ^{59}Co, ^{74}Ge, ^{81}Br, ^{87}Rb, ^{88}Sr, ^{127}I, ^{140}Ce	^{35}Cl, ^{65}Cu, ^{70}Ge, ^{96}Ru, ^{110}Pd, ^{128}Te, ^{160}Gd	^{65}Cu
10^{-4}–10^{-3}	^{48}Ti, ^{61}Ni, ^{69}Ga, ^{72}Ge, ^{73}Ge, ^{77}Ge, ^{94}Zr, ^{101}Ru, ^{127}I, ^{138}Ba, ^{139}La, ^{142}Ce, ^{205}Tl	^{51}V, ^{68}Zn, ^{75}As, ^{90}Zr, ^{110}Pd, ^{115}In, ^{152}Sm, ^{208}Pb, ^{209}Bi	^{154}Sm	^{27}Al, ^{50}Ti, ^{104}Ru, ^{127}I, ^{160}Gd, ^{205}Tl
10^{-3}–10^{-2}	^{109}Ag, ^{140}Ce, ^{142}Nd, ^{153}Eu, ^{187}Re, ^{195}Pt, ^{200}Hg, ^{210}Hg, ^{208}Pb, ^{208}Bi	^{79}Br, ^{94}Zr, ^{108}Pd, ^{130}Te, ^{133}Cs, ^{142}Ce, ^{202}Hg, ^{203}Tl	^{58}Ni	^{23}Na, ^{26}Mg, ^{63}Cu, ^{71}Ga, ^{81}Br, ^{138}Ba, ^{208}Pb, ^{141}Pr
10^{-2}–10^{-1}	^{193}Ir, ^{194}Pt	^{146}Nd, ^{158}Gd, ^{162}Dy, ^{191}Ir, ^{197}Au		^{30}Si, ^{96}Zr, ^{139}La, ^{142}Ce

gross-element analysis. These methods have greatly extended both the scope of charged particle techniques and the range of problems to which they can be applied [24].

Figure 7.2 summarizes the major analytical uses that have been found

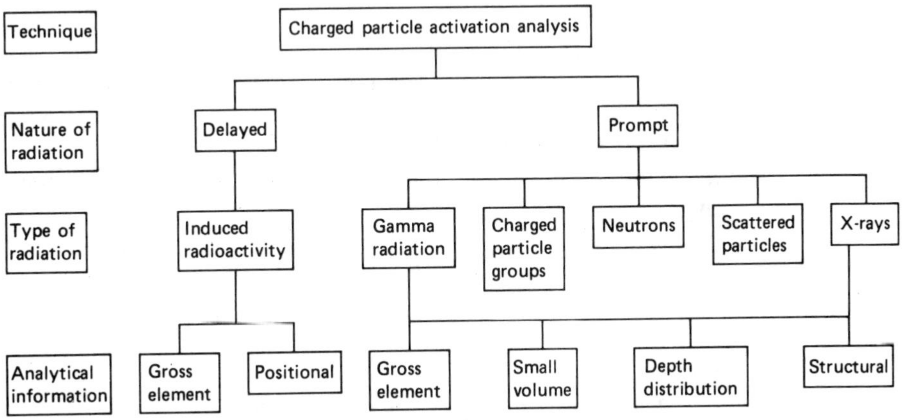

Fig. 7.2 — The major analytical uses of charged-particle activation analysis. (Reproduced from [25] by permission of the copyright holders, Akadémiai Kiadó, Budapest).

for charged-particle techniques, and indicates the overall versatility of the discipline [25]. The methods have been divided into two major categories, one based on the measurement of prompt radiation, which can be measured only during irradiation, and the other on the delayed radiation produced by radioactive decay. Prompt radiations have been subdivided according to type of radiation, and the analytical information that can be obtained is given in the bottom line of the figure.

Charged particles gain the required energy and intensity in accelerators (Cockroft-Walton, van de Graaf accelerators, cyclotrons). Nuclear reactors can also be used to obtain secondary charged particles. Certain compounds are bombarded with neutrons and this leads to the emission of charged particles. The most commonly used reactions are as follows:

$^{6}Li(n,t)^{4}He$ — which represents a source of alpha-particles and tritons,
$^{10}B(n,\alpha)^{7}Li$ — which yields alpha-particles.

However, fluxes of charged particles obtained in this way are usually lower by factors of 10^3–10^9 than those gained in accelerators. This causes a significant drop in sensitivity.

The maximum energy and ion current in the accelerators used in activation analysis can vary. Thus, there are differences between sensitivity values presented by individual authors. In general, Be, B, C, N, O, F and S can be determined when present in concentrations below 10 μg/g, and for Na, Al, Si, P, Cl, and Ca, the concentration limit is below 100 ng/g, when the charged particle energy and ion current are as high as 20 MeV and 50 μA, respectively.

Types and examples of activation with charged particles are given below.

Proton activation. Determination of carbon — $^{12}C(p,n)^{13}N$; oxygen — $^{18}O(p,n)^{18}F$; boron — $^{10}B(p,n)^{11}C$; lithium — $^{7}Li(p,\alpha)^{4}He$.

Deuteron activation. Determination of sodium — $^{23}Na(d,p)^{24}Na$; phosphorus — $^{31}P(d,p)^{32}P$; gallium — $^{69}Ga(d,p)^{70}Ga$; sulphur — $^{32}S(d,\alpha)^{30}P$; nitrogen — $^{14}N(d,n)^{15}O$; carbon — $^{12}C(d,n)^{13}N$; iron — $^{54}Fe(d,n)^{55}Co$; boron — $^{10}B(d,n)^{11}C$.

^{3}He activation. This activation is used mainly for the determination of oxygen:

$$^{16}O(^{3}He,p)^{18}F$$

$$^{16}O(^{3}He,n)^{18}Ne \xrightarrow[1.6\ \text{sec}]{\beta^+} {}^{18}F\ (t_{\frac{1}{2}} = 112\ \text{min})$$

and carbon:

$$^{12}C(^{3}He,\alpha)^{11}C \quad \text{and} \quad ^{12}C(^{3}He,d)^{13}N.$$

In Table 7.4, sensitivity data are summarized for light helium activation

Table 7.4 — Sensitivities of ^{3}He activation analysis

	Sensitivity	
	μg/g	μg/cm^2
B	0.05	0.002
C	0.3	0.01
N	1.0	—
O	0.5	0.01
F	0.5	0.1

analysis, assuming an energy of 5 MeV, beam current of 0.5 μA and an irradiation time of 10 min.

Activation with alpha particles. Determination of oxygen — $^{16}O(\alpha,pn)^{18}F$; $^{16}O(\alpha,d)^{18}F$; carbon — $^{12}C(\alpha,\alpha n)^{11}C$.

Triton activation. The analysis sample is irradiated with neutrons in a reactor together with lithium. The tritons emitted from the reaction $^{6}Li(n,\alpha)^{3}H$ react with oxygen $^{16}O(t,n)^{18}F$. The activity of ^{18}F is measured.

7.2.4 Photon activation analysis

Two principles may be utilized in photon activation analysis [2].

(1) Nuclear photo-effect (photonuclear reaction). Irradiation with gamma-photons releases neutrons, protons, or other particles from the nucleus; e.g. reactions such as (γ,n), (γ,p), $(\gamma,2n)$, $(\gamma,2p)$, $(\gamma,n\beta)$, (γ,np), $(\gamma,{}^{3}He)$, (γ,t), (γ,α).
(2) In elastic gamma-photon scattering, which excites a nucleus to a metastable state (nuclear isomer); reaction (γ,γ').

Intensive gamma-photon fluxes may be obtained from linear accelerators, betatrons and synchrotrons. In certain special cases (determination of beryllium and deuterium) radionuclide gamma-photon sources are sufficient; ^{24}Na, ^{124}Sb, and Ra with its decay products are most often used.

Environmentally significant elements for which the detection limits in thermal neutron activation analysis are relatively high (e.g. Cd, Hg, Pb) can be determined by photon activation with a very low detection limit. The radioactivity of the resulting radionuclides is measured by both conventional gamma-spectrometers and low-energy photon-spectrometers. Table 7.5 lists photonuclear reactions used for instrumental environmental multielement analysis [26].

Photon activation analysis is a useful tool for environmental multielement analysis, especially when applied in connection with low-energy photon-spectroscopy [26–31].

Table 7.5 — Photonuclear reactions used for instrumental multielement analysis [26]. (Reproduced by permission of the copyright holders, Akadémiài Kiadó, Budapest).

Element	Reaction	Half-life	E, keV (θ, %)	Sensitivity (μg/g)
Ni	$^{58}Ni(\gamma,n)^{57}Ni$	36 hr	127 (15)	0.5
			1379 (85)	0.5
	$^{58}Ni(\gamma,p)^{57}Co$	270 days	122 (86)	0.4
Cu	$^{65}Cu(\gamma,n)^{64}Cu$	13.7 hr	7.48 (14)	0.3
Zn	$^{68}Zn(\gamma,p)^{67}Cu$	61.9 hr	93 (16)	0.7
			185 (45)	0.2
As	$^{75}As(\gamma,n)^{74}As$	17.77 days	596 (60)	0.2
Se	$^{76}Se(\gamma,n)^{75}Se$	120 days	136 (58)	0.2
Mo	$^{100}Mo(\gamma,n\beta)^{99m}Tc$	66 hr	141 (85)	0.1
Ag	$^{107}Ag(\gamma,2n)^{105}Ag$	41.2 days	21.18 (60)	0.03
			314 (42)	0.05
Cd	$^{116}Cd(\gamma,n\beta)^{115m}In$	53.38 hr	336 (45)	0.01
Sn	$^{112}Sn(\gamma,p)^{111}In$	2.83 days	245 (94)	0.1
	$^{118}Sn(\gamma,n)^{117m}Sn$	14 days	158 (65)	1.0
Hg	$^{198}Hg(\gamma,n)^{197}Hg$	64.1 hr	68.81 (36)	0.002
Tl	$^{203}Tl(\gamma,n)^{202}Tl$	12.2 daus	70.82 (42)	0.05
			40 (95)	0.1
Pb	$^{204}Pb(\gamma,n)^{203}Pb$	52.1 hr	72.87 (47	0.4
			279 (81)	0.5
U	$^{238}U(\gamma,n)^{237}U$	6.75 days	59.54 (35)	0.005
			97.08 (18)	0.02
			101.07 (27)	0.01
			208 (23)	0.02

7.3 INSTRUMENTAL ACTIVATION ANALYSIS

Instrumental activation analysis is the method most often used in activation analysis in environmental studies.

For trace element determination in solid environmental samples (air dust, aerosols, plant material, soil samples, ash, sludge, sediments) instrumental neutron activation analysis (INAA) is extremely useful, for the following reasons [15].

(1) Owing to the low detection limits small samples are sufficient.
(2) Activation of the main constituents of this kind of sample is negligible.
(3) Intermediate steps such as dissolution and separation are not necessary.

Figure 7.3 gives an example of INAA of fly ash, by use of two different irradiation times. Gamma-spectra were measured at various time intervals after irradiation and in these spectra the long-lived radionuclides gradually become dominant [21].

For the analysis of samples of biological origin, IAA has only a limited use. The highly active matrix nuclides (^{24}Na, ^{42}K, ^{82}Br, ^{32}P) interfere with the determination of elements which form radionuclides in the sample, e.g. Cu, As, Mo, Cd, etc. On the other hand, elements which form long-lived radionuclides (e.g. Zn, Sb, Co, Fe, Cr, Hg, Se) can often be determined [21].

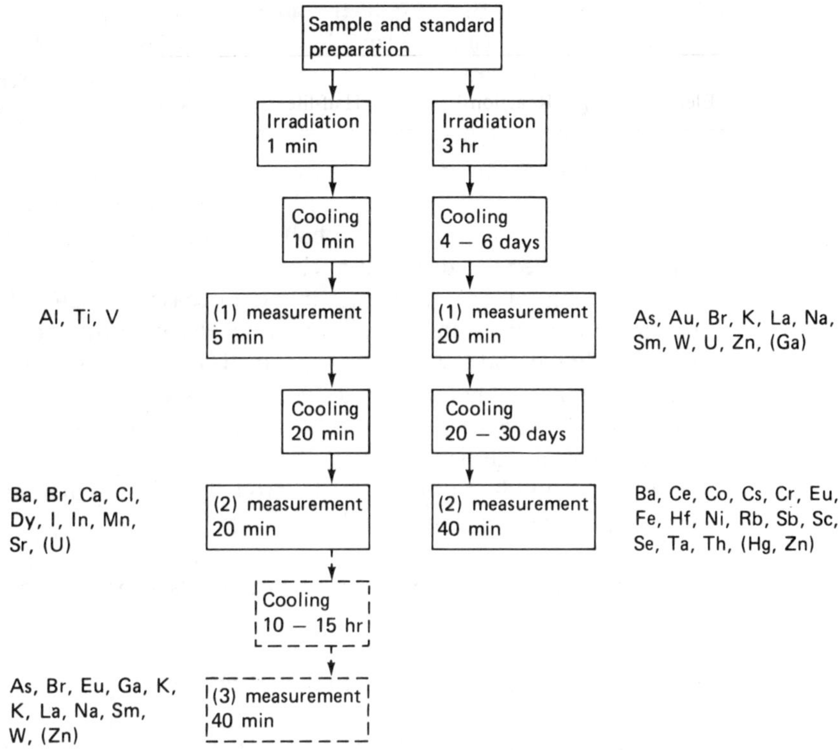

Fig. 7.3 — Procedure for analysis of fly ash with instrumental neutron activation analysis. (Reproduced from [21] by permission of the copyright holders, SNTL, Prague).

The introduction of high-resolution Ge(Li) and Si(Li) detectors for gamma- and X-ray spectrometry has made possible the use of short-lived radionuclides in IAA. Complex spectra can be resolved without the need for chemical separations. The application of cyclic activation-counting enables the use of radionuclides with half-lives shorter than 10 sec.

The advantages of using short-lived radionuclides in activation analysis are [2,21,32,33] as follows.

(1) The sensitivity is enhanced for a given flux of particles and a given cross-section, and the saturation factor $[1-\exp(-\lambda t)]$ can approach the maximum value of unity after a relatively short irradiation time, typically one half-life. Very short-lived activities allow further enhancement of the detection sensitivity by the use of high-intensity reactor pulses, if a pulsing reactor is available.
(2) The analysis is rapid (total analysis time, including irradiation, decay and counting, only about 3–5 times the half-life of the species measured).
(3) The increased speed of the analysis decreases the cost per analysis.

(4) The time between the start of the analysis and the availability of the results can be very short (from a few hours to a few minutes). Thus the use of short-lived radionuclides can make activation analysis more competitive with other methods of analysis, even in routine analysis for control of environmental quality.
(5) The selectivity of the analysis is increased by the preferential formation of the short-lived radionuclides, and during the measurement not only the gamma-energy but also the half-life can be used as a means of selection between the different nuclides.
(6) The precision of the analysis can be increased because in purely instrumental analysis the sample can, after decay of the radionuclides, be re-irradiated and the analysis repeateed. In such cases the sample is also quickly available for other kinds of analysis.
(7) There is no need to seal samples in fused silica if only short-lived radionuclides are utilized.
(8) There is less build-up of long-lived matrix activities, so handling, waste disposal, etc. are all simplified.

Along with such advantages, certain disadvantages must also be recognized [33].

(1) For a given initial photo-peak counting rate, fewer photo-peak counts are obtained during counting.
(2) The half-lives are too short for effective post-irradiation radiochemical separations for the removal of interfering activities.
(3) A high-speed, automated transfer system is needed.
(4) Container blank problems may be difficult to avoid in some cases.
(5) Difficulties with high initial counting dead-times, which change rapidly in a complex fashion, are often encountered.

For a number of elements (Mn, Ga, As, Se, Ru, La, Eu, Ho, Lu, Os and Au), INAA by use of radionuclides with moderate life-times (2 hr–3 days) is the most sensitive, even after a 1-min irradiation and immediate counting. For all other elements the highest gamma emission rate results from nuclides with half-lives shorter than 1 hr. Extremely high counting rates are obtained after thermal neutron activation for Na, Sc, V, Co, Se, Rh, Ag, In, Eu, Dy, Er and Hf. For some elements (F, Si, P, Cl, Ca, Cr, Se, Y, Ba, Ce, Pr, Bi, Pb), activation with reactor fission neutrons, or 14 MeV neutrons, provides an interesting sensitivity for a threshold reaction or the production of an isomer. Cyclic activation and counting of very short-lived radionuclides has been applied advantageously in the ng–μg range for a number of elements [32].

Low-energy photon activation analysis allows selective production of metastable isomers of Se, Br, Ag, Er, Hf, Ir and Au, while high-energy photon analysis yields high-specific activities for elements such as K, Cr, Se, Br, Mo, Pr, Nd, Ho, and allows interesting determinations of Mg, Cl, Si, Cr, Fe, Zr and Pb. Charged-particle activation can be used with high sensitivity for some medium and high-Z elements (Cr, Y, Se, Br, Zr, Mo, La, Ta, V) [32].

7.4 DESTRUCTIVE (RADIOCHEMICAL) ACTIVATION ANALYSIS

In destructive activation analysis post-irradiation chemical separations are used. An inactive carrier of the element is added as early as possible in the chemical procedure. The sample should be completely dissolved and steps taken to ensure isotopic exchange between radioactive and carrier species of the element. Separation methods known from classical analytical chemistry (such as precipitation, extraction, electrolysis, chromatography), as well as special radiochemical methods (such as evaporation, distillation, isotopic exchange, electrophoresis, polarography) are used. Separation methods are chosen that provide a selective separation of the radionuclide of interest. Some of these separations can be repeated several times and individual separation steps need not be quantitative. Ultimately, the chemical recovery of the element and its carrier must be established by chemical techniques. Substoichiometric isolation avoids the need for chemical recovery determinations. The automation of radiochemical separation procedures facilitates the analysis of large numbers of samples [1,2,34–37].

The destructive method is used mainly for analysis of biological materials. Figure 7.4 shows an example of a procedure for radiochemical neutron activation analysis of human liver [34].

The disadvantages of radiochemical activation analysis are contact with highly active samples and lengthy procedures, which make the use of short-lived radionuclides practically impossible.

7.5 ANALYSIS OF ENVIRONMENTAL SAMPLES

7.5.1 Water

Water samples may be irradiated without pretreatment, but usually they are evaporated to dryness and the residue is irradiated. Volatile constituents (e.g. Hg, As) may be lost during drying. The mildest method is freeze-drying. Several 100-ml portions may be dried on a polyethylene foil and thus reduced to a very small volume for irradiation. The enrichment factor obtained by drying depends on the salt content of the water and is of the order of 10^2–10^4 [15].

Determination of trace element concentrations by direct irradiation in a high salt matrix such as sea-water, estuarine water, or brine, with thermal neutrons, is particularly difficult owing to severe interferences from intense activities of major nuclides, such as ^{24}Na, ^{38}Cl and ^{82}Br. The high levels of Na, Cl and Br produce an extremely high background level of radiation that totally obscures the signals of most elements yielding neutron activation products which have comparable half-lives. Therefore, effective separation of sodium chloride from the other constituents is necessary. Hydrated

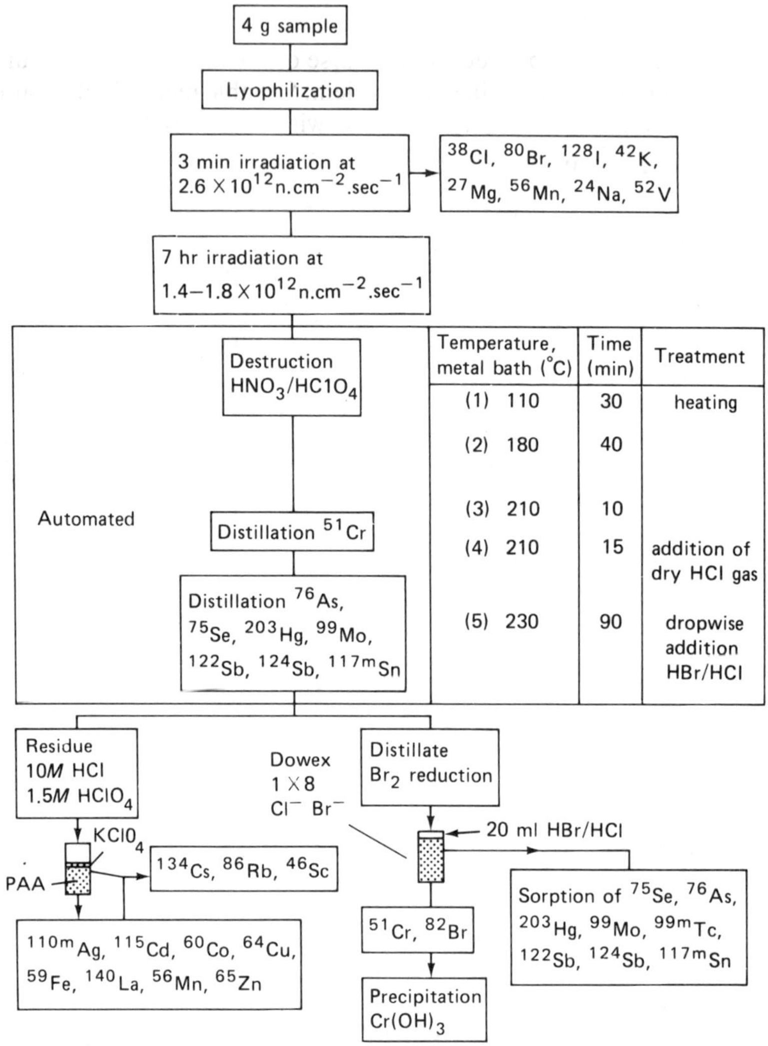

Fig. 7.4 — Separation scheme for trace elements in human liver [34]. (Reproduced by permission of the copyright holders, Akadémiai Kiadó, Budapest).

antimony pentoxide (HAP) is an excellent scavenger of Na^+ from strong acid solutions, with a retention capacity of about 31 mg of Na^+ per g of HAP [35,36,38,39]. The main advantage of a post-irradiation separation is the elimination of contamination. One of the disadvantages of the post-irradiation separation of ^{24}Na is that only nuclides with half-lives considerably longer than that of ^{24}Na can be utilized for elemental determination, unless a hot-cell is available for processing the highly radioactive sample immediately after the end of the irradiation. If handling of the very high activities of sodium is to be avoided, the trace elements and sodium chloride must be separated before irradiation. This is unfavourable, because one of the main advantages of activation analysis, namely its insensitivity to impurities introduced in the course of the analysis, is lost, but there are no other means of solving the problem. Suitable methods of separation must be selected and the separation done with great care [15, 40, 41].

Table 7.6 is a review of recent work dealing with water analysis.

Table 7.6 — Water analysis

Matrix	Elements determined	References
Surface water and sea-water	V	[42]
Tap, fresh, surface and sea water	Se	[43]
Waste water	Hg	[44]
Natural water	Cs	[45]
Rain, tap, surface, sea water	Br, I	[46]
Snow	Cd, Hg	[47]
Surface water	Co, Ag	[48]
Surface water	V, Mn, Cu	[49]
River water	Cd, As, Zn	[50]
River water and sea-water	Cr, Fe, Co, Zn, Mo, Ag, Sb, Au, U	[51]
Sea-water	Cr, Co, Zn, Ag, Cd, Hg	[52]
Sea-water	Fe, Co, Ni, Zn, Ag, Sb, Au, Hg	[53]
River water	Multielement analysis	[54–56]
Sea-water	Multielement analysis	[55]
Fresh water	Multielement analysis	[57]
Geothermal water	Multielement analysis	[58]
Treated water	Multielement analysis	[59]
Atmospheric water	Multielement analysis	[60, 61]

7.5.2 Air particulate matter

Sampling is the most important and difficult step in the analysis of air particulates. If the sample collected is not representative, the final results will be of little or no value. Methods of collection of air particulates are presented in Section 2.1. The packed air particulate samples are enclosed in irradiation cans or 'rabbits' and irradiated.

Although activation analysis of particulate matter usually requires no treatment of the samples before radioactivity measurement, in some cases (e.g. determination of certain trace elements producing pure beta-emitters) it is necessary to dissolve the sample. Wet ashing with acids is frequently a suitable procedure, although dry ashing may also be used. If wet ashing is done in a distillation unit, volatile bromides of As, Hg, Sb, Se and Sn may be distilled after the addition of hydrobromic acid. The other elements may then be separated as groups or individual elements by various methods (precipitation, co-precipitation, extraction, chromatrography, ion-exchange, electro-deposition, electrophoresis, etc.) [62].

Very often group separation is used in analysis of air dust samples [63].

The non-volatile trace elements in sulphuric acid solution are separated, after removal of silica as SiF_4, from alkali metals , alkaline-earth metals and phosphorus by a selective sorption step on an ion-exchange column (Bio-Rad Chelex 100). The separation is performed at pH ~ 4 after buffering with dilute sodium acetate, so that a large number of multivalent cations are chelated in the resin bed. In this way the elements can be fractionated into four groups that can be analysed by using a Ge(Li) detector and a multichannel analyser [62].

Activation analysis is used for the determination of trace elements in air particulates. Some authors have dealt with activation analysis of air particulates for one element or a limited number of elements that are usually difficult to determine by conventional methods. Most of these works are devoted to multielement analysis. The main objectives of research by use of activation analysis are the investigation of particles from major air-pollution sources, analysis of effluents from coal-fired power plants, and investigation of the continental, marine and man-made contribution to the composition of aerosols (particularly in industrial regions), interpretation of aerosol trace elements, estimation of the levels of the hazardous trace elements and their dependence on the concentration of major pollutants, as well as the dependence of trace element concentrations on meteorological parameters. For example, the distribution of halogens in atmospheric air [64–66], the superimposed contribution of industrial stack effluents to the natural aerosol composition [67], the interrelations between inorganic and organic compounds present in the environment and in the organism, with particular attention to the evaluation of possible hazards [68], the atmospheric contamination of Mexico City [69,70], Japanese cities [71–74], Hong-Kong [75] and Caracas [76], have been studied by use of activation analysis.

An excellent review on activation analysis of air particulate matter has been published by Alian and Sensoni [62].

Table 7.7 is a review of recent work dealing with the analysis of air particulate matter.

7.5.3 Plants, soils and sediments

Plants can be affected directly by air pollutants, as well as indirectly through the contamination of soil and water. At the same time, plants are a member of the food chain and may create a risk for man and animals through contamination of food supplies.

In highly polluted areas, certain plants show increased concentration of elements specific for impurities in air, water and soil, and thus can serve as a long-term monitor of environmental pollution. An advantage of these bio-indictors is that they are stationary, commonly available in large numbers and have relatively equal response [132].

For activation analysis, samples of plant material are prepared by dry and wet ashing, or by pulverizing dried plants.

Table 7.7 — Air particulate matter analysis

Matrix	Elements determined	References
Aerosols	Se	[77]
Aerosols	In	[78]
Flue dust	Re	[79]
Polluted atmosphere	Si	[80]
Polluted atmosphere	Hg	[80,81]
Polluted atmosphere	Ag	[82]
Polluted atmosphere	Zn	[83]
Polluted atmosphere	U	[84]
Polluted atmosphere	S	[85]
Fall-out	Hg	[86]
Aerosols	Br	[87]
Aerosols	Al, Si	[88]
Aerosols	V, Zn	[89]
Aerosols	Cl, Br, I	[64]
Aerosols	Br, Mn, Na	[90]
Polluted atmosphere	Br, Cd, Pb	[91–93]
Polluted atmosphere	Br, Pb, V	[94]
Polluted atmosphere	Si, Rb, Y	[95]
Fly ash	V, W, Yb, Zn	[96]
Airborne gunshot residues	Sb, V, W, Zn	[97]
Atmosphere of underground U-mine	Cu, Fe, Pb, Sb, Sc, Tc, V, Zn	[98]
Air dust	multielement analysis	[99–103]
Aerosols	multielement analysis	[104–117]
Surface air	multielement analysis	[118–120]
Fly ash	multielement analysis	[111,121–124]
Pacific marine aerosol	multielement analysis	[125]
South Pole aerosol	multielement analysis	[126]
Particulates emitted from aircraft jet engines	multielement analysis	[127]
Automobile exhaust pollutants	multielement analysis	[128]
Polluted atmosphere	multielement analysis	[129]
Marine atmosphere	multielement analysis	[130,131]

Table 7.8 is a review of recent work dealing with plants, soil and sediment analysis.

7.5.4 Biological materials

Activation analysis of biological materials is becoming more and more important. This is associated with the fact that many tasks solved with the aid of activation analysis (such as analyses of one hair, one muscle fibre, a bioptic sample of kidney tissue) could not be solved by using other methods.

Activation analysis is an important technique for analysing various human and animal tissues (nerve fibres, muscle tissue, bones, teeth, hairs

Table 7.8 — Plants, soils and sediments analysis

Matrix	Elements determined	References
Standards: NBS Spinach, NBS Tomato leaves, NBS Pine needles, NIES Pepperbush	P	[133]
Standards: Bowen's Kale, NBS Orchard leaves, NBS Spinach, IAEA Potatoes, etc.	V	[134]
Standards: NBS Orchard leaves, NBS Pine needles, NBS Spinach, NBS Tomato leaves, Soils, Sea sediments	I	[136–138]
Standards: NBS Orchard leaves, Bowen's Kale, Lake water sediment	Sb	[139]
Standards: IAEA Wheat flour, IAEA Dried potatoes, IAEA Powdered milk	Mo	[140]
Standard: Bowen's Kale	Hg	[141]
Plants, soil, standards	U	[142]
Lake sediments	As	[143]
Grass and clover flowers	P, N	[144]
Wheat, soil	As, Sb	[145]
Standards: Bowen's Kale, NBS Orchard leaves	Au, As	[146]
Standards: NBS Orchard leaves, NBS Pine needles, NBS Tomato leaves	Cl, Br, I	[147]
Standards: NBS Spinach, NBS Orchard leaves	In, Cu, Ni	[148, 149]
Standards: NBS Orchard leaves, NBS River sediments	Cd, Cu, Hg	[150]
Pepperbush	Co, Ni, Rb, Sr	[151]
Standards: Bowen's Kale, NBS Orchard leaves	Cu, Zn, Cd, As, Sb	[153]
Standards: NBS Orchard leaves, NBS Spinach, Green and black tea	Mn, Cr, Co, Ni, Zn, Pb	[153]
Tobacco	Sc, Cr, Fe, Co, Sb, La	[154]
Standards: NBS Orchard leaves, Bowen's Kale	Multielement analysis	[155]
Tea leaves	Multielement analysis	[156, 157]
Tobacco	Multielement analysis	[158-162]
Plants	Multielement analysis	[163, 164]
Rice	Multielement analysis	[165]
Spices	Multielement analysis	[166]
Standards: Bowen's Kale, IAEA Powdered milk	Multielement analysis	[167]
Standards: NBS Orchard leaves, IAEA Wheat flour	Multielement analysis	[168]
Standards: Bowen's Kale, IAEA Dried potatoes	Multielement analysis	[169]
Standards: NBS Orchard leaves, Bowen's Kale	Multielement analysis	[170]
Standards: NBS Orchard leaves, NBS Spinach, NBS Tomato leaves, NBS Pine needles	Multielement analysis	[171]
Standards: Drug plants, soil	Multielement analysis	[172]
Lake sediments	Multielement analysis	[173]
River sediments	Multielement analysis	[150, 174, 175]
Sea sediments	Multielement analysis	[176, 177]
Soils	Multielement analysis	[178, 179]
Sludges	Multielement analysis	[178, 180]

and marine organisms) and in analysing human as well as animal fluids, such as the blood (blood plasma, erythrocytes), urine, bile and cerebrospinal fluid. The greatest attention has been paid to blood and its components.

Activation analysis is normally applied *in vitro* to small samples of the material of interest, but it is also applicable *in vivo* to the whole body of a living subject or parts of the body.

Samples of biological materials are suitable for neutron activation analysis, because the elements in high concentrations (H, C, N, O) are activated by thermal neutrons only to a small extent and the self-absorption of neutrons in these materials is very small. Elements in low concentrations, such as Na, Cl, K, Br, P, complicate the determination of trace elements which produce radionuclides with half-lives from 20 min to 50 hr. With instrumental neutron activation analysis it is sometimes possible to determine elements which produce short-lived radionuclides (^{77m}Se or ^{52}V with half-lives of 17.5 sec and 3.75 min respectively) or in contrast long-lived nuclides (^{203}Hg, ^{60}Co, ^{59}Fe, ^{51}Cr, ^{65}Zn and ^{124}Sb). A review of the elements which can be determined in biological materials by INAA is given in Fig. 7.5 [21].

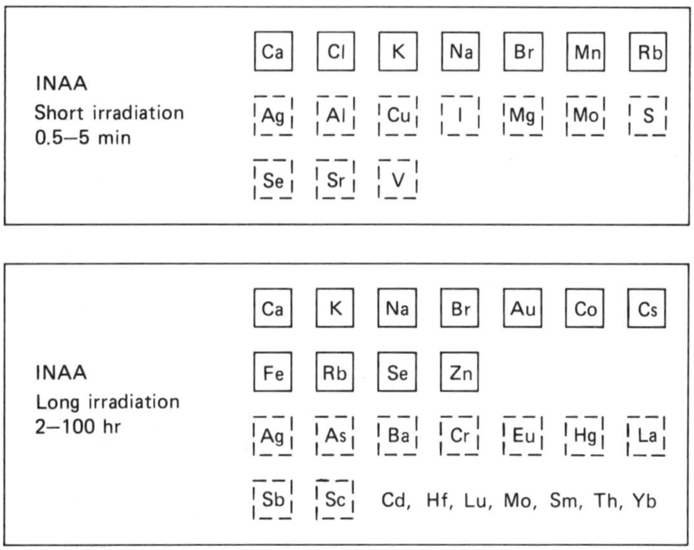

Fig. 7.5 — Elements determinable in biological materials by INAA.
[X] elements which can be determined in all types of biological materials,
[X] (dashed) elements which can be determined only in some cases,
X elements which can only rarely be determined. (Reproduced by permission of the copyright holders, SNTL, Prague).

For the determination of important trace elements, such as As, Cd, Cr, Cu, Hg, Mo, Ni and Se, in biological materials, radiochemical activation analysis usually has to be used [35, 181, 182].

Table 7.9 is a review of recent work dealing with the analysis of biological materials.

Table 7.9 — Analysis of biological materials

Matrix	Elements determined	References
Biological materials	Ba	[183]
Milk	I	[184–186]
Blood and blood serum	I	[184]
Standards: NBS Bovine liver, BCR Milk	I	[185]
Standards: NBS Oyster tissue, NBS Bovine liver, IAEA Oyster tissue, IAEA Blood, IAEA Milk powder	I	[187]
Standards: NBS Oyster tissue	Na	[188]
Standards: NBS Bovine liver, IAEA Animal muscle tissue, Sea biota, Human blood serum, plasma	V	[134, 189–191]
Standard: NBS Oyster homogenate	Ni	[192]
Oyster homogenate, animal muscles	Co	[193]
Standards: NBS Oyster tissue, NBS Bovine liver, Human liver, Human and animal blood serum	Se	[194–196]
Standards: NBS Bovine liver, IAEA Animal muscle, River fish, Human nail and hair	As	[197–202]
Standards: NBS Bovine liver, Rat liver, Rat kidney, Rat spleen	Zr	[203]
Human blood serum, liver, lung, kidney	Ga	[204]
Hair	Hg	[205,206]
Tooth enamel	F	[207]
Kidney	Cd	[208,209]
Hair	As, Hg	[210]
Sea fish	K, Cs	[211]
Standards: NBS Bovine liver, Human liver specimens	Au, Pt	[212]
Human blood serum	I, Br	[213]
Animal blood	Br, As	[214]
Standards: NBS Bovine liver, Fish, Human encephalon, lung, liver kidney	As, Cd, Hg	[215–217]
Standards: NBS Bovine liver, A-2 Animal whole blood, A-3/1 Calcined animal bone	As, Se, Sb	[218]
Standard: NBS Bovine liver	V, Mo, As	[219]
Chromatin of lymphocytes	Cu, Zn, Mn	[220]
Fish	Mn, Zn, Cr, Ag	[221]
Standards: NBS Bovine liver, Human milk	Cu, Zn, Mn, As, Mo	[223]
Beef extract	As, Cd, Cu, Sn, Zn	[223]
Milk formulas, powder milk, infant food	Fe, Zn, Co, Cr, Rb, Cu, As	[224]
Milk powder	Multielement analysis	[225–227]
Brewer's yeast	Multielement analysis	[228]
Sea biota	Multielement analysis	[229]
Standards: Fish, Animal muscles	Multielement analysis	[171]
Standard: NBS Bovine liver	Multielement analysis	[230]
Animal liver, kidney and muscle	Multielement analysis	[231–232]
Monkey liver, kidney and spleen	Multielement analysis	[233]
Human kidney and lung	Multielement analysis	[234,235]
Standards: NBS Bovine liver, IAEA Animal muscle, IAEA Animal blood, Human blood serum and plasma	Multielement analysis	[236–239]
Human blood	Multielement analysis	[239–241]
Human urine, bone ash	Multielement analysis	[240]
Hair	Multielement analysis	[242–247]

REFERENCES

[1] J. Tölgyessy and Š. Varga, *Nuclear Analytical Chemistry*, Vol. III, *Radiochemical and Activation Analysis*, VSAV, Bratislava; University Park Press, Baltimore, 1974.

[2] J. Tölgyessy and M. Kyrš, *Radioanalytical Chemistry*, VEDA, Bratislava; E. Horwood, Chichester, in the press.
[3] J. Tölgyessy, Yu. V. Yakovlev and G. N. Bilimovich, *Diagnostika okruzhayushche sredy radioanaliticheskimi metodami*, Energoatomizdat, Moscow, 1985.
[4] M. Rakovič, *Activation Analysis*, Academia, Prague, 1970.
[5] P. Kruger, *Principles of Activation Analysis*, Wiley–Interscience, New York, 1971.
[6] D. De Soete, R. Gijbels and J. Hoste, *Neutron Activation Analysis,* Wiley–Interscience, London, 1972.
[7] S. Amiel (ed.), *Nondestructive Analysis with Nuclear Reactors and Radioactive Neutron Sources*, Elsevier, Amsterdam, 1981.
[8] J. M. A. Lenihan and S. J. Thomson, *Advances in Activation Analysis*, Vol.1, Academic Press, London, 1969.
[9] G. Deconninck, G. Demortier and F. Bodart (eds.), *Chemical Analysis by Charged Particle Bombardment*, Elsevier, Lausanne; Akadémiai Kiadó, Budapest, 1972.
[10] R. Kuznetsov, *Aktivatsionnyi analiz*, Atomizdat, Moscow, 1967.
[11] W. S. Lyon, *Physical Techniques of Activation Analysis*, Elsevier, Amsderdam, 1967.
[12] S. A. Nargolwalla and E. P. Przybylowicz, *Activation Analysis with Neutron Generators*, Wiley, New York, 1973.
[13] E. Szabó and A. Simonits, *Aktivációs analízis*, Müszaki Könyvkiadó, Budapest, 1973.
[14] A. Alian and B. Sansoni, *J. Radioanal. Nucl. Chem., Articles*, 1985, **89**, 191.
[15] K. H. Lieser, *Nuclear Analytical Techniques and their Application to Environmental Samples*, Consultants' Meeting on Advances in Nuclear Techniques for Analysis of Environmental Samples, Karlsruhe, FRG, 19–21 February 1986.
[16] A. J. Moses, *Nuclear Techniques in Analytical Chemistry*, Pergamon Press, Oxford, 1964.
[17] Z. E. Pelekis, in *Yadernaya spektroskopiya i neitronnoaktivatsionnii analiz*, p. 31. Izd. Zinatne, Riga, 1965.
[18] A. Simonits, F. De Corte and J. Hoste, *J. Radioanal. Chem.*, 1975, **24**, 31.
[19] Z. Řanda, M. Vobecký, J. Kunciř and J. Benada, *J. Radioanal. Chem.*, 1978, **46**, 95.
[20] R. A. Nadkarni and G. H. Morrison, *J. Radioanal. Chem.*, 1978, **43**, 347.
[21] I. Obrusník, *Neutronová aktivačná analýza*, in *Nové směry v analytické chemii*, J. Zýka (ed.), Vol. II, p. 163, SNTL, Prague, 1984.
[22] L. Moens, F. De Corte, J. Hoste and A. Simonits, *J. Radioanal. Chem.*, 1978, **45**, 221.
[23] L. Gy. Nagy and L. Szokolyi, *Neutronaktiválásos vizsgálatok*, Müszaki Könyvkiadó, Budapest, 1966.
[24] Chemical Analysis by Charged Particle Bombardment, Intern. Meeting, Namur (Belgium) 1971; *J. Radioanal. Chem.*, 1972, **12**.
[25] T. B. Pierce, *J. Radioanal. Chem.*, 1972, **12**, 23.
[26] Chr. Segebade, H.-U. Fusban and H.-P. Weise, *J. Radioanal. Chem.*, 1980, **59**, 399.
[27] J. S. Hislop, *Photon Activation Analysis of Biological and Environmental Samples: A Review. Proc. Intern. Conf. on Photonuclear Reactions and Applications.* Asilomar Conference Grounds, Pacific Grove, California, USA, 26–30 March, 1973.
[28] N. K. Aras, W. H. Zoller, G. E. Gordon and G. J. Lutz, *Anal. Chem.*, 1973, **45**, 1481.
[29] A. Chattopadhyay and R. E. Jervis, *Anal. Chem.*, 1974, **46**, 1630.
[30] G. J. Lutz, *J. Radioanal. Chem.*, 1974, **19**, 139.
[31] J. J. Paciga and R. E. Jervis, *Environ. Sci. Technol.*, 1976, **10**, 1124.
[32] R. Dams, *J. Radioanal. Chem..*, 1981, **61**, 13.
[33] V. P. Guinn and D. A. Miller, *J. Radioanal. Chem.*, 1977, **37**, 313.
[34] P. Lievens, J. Versieck, R. Cornelis and J. Hoste, *J. Radioanal. Chem.*, 1977, **37**, 483.
[35] F. Girardi and R. Pietra, *At. Energy Rev.*, 1976, **14**, 521.
[36] F. Girardi and E. Sabbioni, *J. Radioanal. Chem.*, 1968, **1**, 169.
[37] R. A. Nadkarni and G. H. Morrison, *Anal. Chem.*, 1975, **47**, 2285.
[38] F. Girardi, R. Pietra and E. Sabbioni, *EURATOM Rept.*, EUR 4287 e, 1969.
[39] F. Girardi, R. Pietra and E. Sabbioni, *J. Radioanal. Chem.*, 1970, **5**, 141.
[40] S. K. Nyarku and A. Chatt, *J. Radioanal. Chem.*, 1982, **71**, 129.
[41] R. R. Greenberg and H. M. Ringston, *J. Radioanal. Chem.*, 1982, **71**, 147.
[42] H. A. van der Sloot and H. A. Das, *J. Radioanal. Chem.*, 1977, **35**, 139.
[43] R. Massée, H. A. van der Sloot and H. A. Das, *J. Radioanal. Chem.*, 1977, **35**, 157.
[44] Houng-Huei Lin, *Radiochem. Radioanal. Lett.*, 1979, **37**, 327.
[45] T. Ganzerli, M. Valentini, R. Stella, L. Maggi and G. Ciceri, *7th Intern. Conf. Modern Trends in Activation Analysis*, 23–27 June 1986, Copenhagen, MTAA-7211.

[46] J. B. Luten, H. A. Das and C. L. De Ligny, *J. Radioanal. Chem.*, 1977, **35**, 147.
[47] E. A. Danilova, L. I. Zhuk and A. A. Kist, in *Yaderno-fizicheskie metody analiza v kontrole okruzhayushchei sredy*, p. 148. Gidrometeorizdat, Leningrad, 1980.
[48] N. Genova, V. C. Crespi and S. Meloni, *Radiochem. Radioanal. Lett.*, 1983, **58**, 271.
[49] O. E. Veveris, E. B. Kalinin, L. L. Pelekis *et al.*, in *Yaderno-fizicheskie metody analiza v kontrole okruzhayushchei sredy*, p. 145. Gidrometeorizdat, Leningrad, 1980.
[50] S. Sava, L. Zikovsky and J. Boisvert, *J. Radioanal. Chem.*, 1980, **57**, 23.
[51] V. A. Vizhenskii, E. I. Andrievski, T. P. Molchanova *et al.*, in *Yaderno-fizicheskie metody analiza v kontrole okruzhayushchei sredy*, p. 211. Gidrometeorizdat, Leningrad, 1980.
[52] O. V. Stepanets, B. Ya. Spivakov, Yu. V. Yakovlev, V. P. Kolotov and B. V. Savel'ev: *Zh. Analit. Khim.*, 1976, **31**, 440.
[53] O. E. Veveris, M. V. Virtsavs, Z. B. Kalkinya *et al.*, in *Yaderno-fizicheskie metody analiza v kontrole okruzhayushchei sredy*, p. 32. Gidrometeorizdat, Leningrad, 1980.
[54] Ch. L. Ndiokwere and V. P. Guinn, *J. Radioanal. Chem.*, 1983, **79**, 147.
[55] C. Lee, N. B. Kim, I. C. Lee and K. S. Chung, *Talanta*, 1977, **24**, 241.
[56] S. Habib and M. J. Minski, *J. Radioanal. Chem.*, 1981, **63**, 379.
[57] A. Wyttenbach, S. Bajo and K. Farrenkothen, *Radiochem. Radioanal. Lett.*, 1980, **42**, 307.
[58] V. A. Luknickii and N. A. Lepechina in *Yaderno-fizicheskie metody analiza v kontrole okruzhayushchei sredy*, p. 135. Gidrometeorizdat, Leningrad, 1980.
[59] N. V. Shtangeeva, V. D. Kulikov, L. V. Mikheeva *et al.*, in *Yaderno-fizicheskie metody analiza v kontrole okruzhayushchei sredy*, p. 218. Gidrometeorizdat, Leningrad, 1980.
[60] P. Schutyser, W. Maenhaut and R. Dams, *Anal. Chim. Acta*, 1978, **100**, 75.
[61] J. B. Luten, *J. Radioanal. Chem.*, 1977, **37**, 897.
[62] A. Alian and B. Sansoni, *J. Radioanal. Nucl. Chem., Articles*, 1985, **89**, 191.
[63] P. Schramek, K. Samsahl and J. Pavlu, *J. Radioanal. Chem.*, 1974, **19**, 329.
[64] R. Lininger, *J. Geophys. Res.*, 1966, **71**, 2457.
[65] R. H. Loucks, *NBS Special Publication 312*, p. 36, Vol. 1, (1968).
[66] J. W. Winchester, *Trans. Am. Nucl. Soc.*, 1965, **8**, 326.
[67] Sř. Tarras and O. L. Pirtle, Jr., *Trans. Am. Nucl. Soc.*, 1962, **5**, 280.
[68] J. Parizek, *Proc. Symp. Nuclear Activation Techniques in the Life Sciences*, Bled, 10–14 April 1972, p. 177. IAEA, Vienna, 1972.
[69] M. Navarrete, L. Galvez, E. Tzontilmatzin and C. Aguilar, *Radiochem. Radioanal. Lett.*, 1974, **19**, 163.
[70] M. Navarrete and L. Galvez, *Proc. Symp. Nuclear Activation Techniques in the Life Sciences*, Bled, 10–14 April 1972, p. 277. IAEA, Vienna, 1972.
[71] A. Aoki, *Kyoto Furitsu Daigaku Gakujutsu Hokoku (Rigaku)*, 1972, No. 23, 55.
[72] T. Kato, N. Sato and N. Suzuki, *Talanta*, 1976, **23**, 517.
[73] T. Mamuro, *Genshiryoku Kogyo*, 1971, **17**, 19.
[74] T. Mamuro, *Proc. Japanese Conf. Radioisotopes*, Aug. 1974, No. 11, p. 125
[75] L. S. Chuang, L. S. Kwong and S. J. Yeh, *J. Radioanal. Chem.*, 1979, **49**, 103.
[76] L. Escalona and E. Sanhueza, *Atom. Environ., 1981,* **15**, 61.
[77] D. Kut and Y. Sarikaya, *J. Radioanal. Chem.*, 1981, **62**, 161.
[78] E. Taskaev and D. Apostolov, *J. Radioanal. Chem.*, 1979, **49**, 127.
[79] A. Govaerts, R. Gijbels and J. Hoste, *Anal. Chim. Acta*, 1972, **62**, 79.
[80] R. van Frieken and R. Dams, *Anal. Chim. Acta*, 1973, **63**, 369.
[81] W. Zmijewska, *J. Radioanal. Chem.*, 1977, **35**, 389.
[82] A. Albini, A. Cesana and M. Terrani, *J. Radioanal. Chem.*, 1976, **34**, 185.
[83] T. Nakanishi ánd B. Sansoni, *AED-Conf-75*, 638–005.
[84] R. Dams, K. A. Rahn, J. A. Robbins, G. D. Nifong and J. W. Winchester, *Proc. 2nd Intern. Clean Air Congress*, p. 509. Academic Press, New York, 1971.
[85] G. Shani and A. Hacoum, *Proc. Symp. Measurement, Detection and Control of Env. Pollution*, Vienna, 15–19 March 1976, p. 89. IAEA Vienna, 1976.
[86] A. Jernelor and T. Wallin, *Atm. Environ.*, 1973, **7**, 2.
[87] G. Shani, Z. Handani and D. Cohen, *IAEA Rept., INIS-mf-121*, 1973, p. 1336.
[88] V. Cercasov, *J. Radioanal. Chem.*, 1979, **52**, 399.
[89] R. C. Ragaini, H. R. Ralston, D. Garvis and R. Kaifer, *UCRL-51850*, 18 June 1975.
[90] C. M. Gordon and R. E. Larson, *J. Geophys. Res.*, 1964, **69**, 2881.
[91] R. A. Duce and J. W. Winchester, *Radiochim. Acta*, 1965, **4**, 100.
[92] R. A. Duce and J. W. Winchester, *J. Geophys. Res.*, 1963, **68**, 3943.
[93] R. A. Duce and J. W. Winchester, *J. Geophys. Res.*, 1965, **70**, 1775.

[94] J. L. Moyers, W. H. Zoller, R. A. Duce and G. L. Hoffmann, *Environ. Sci. Technol.*, 1972, **6**, 68.
[95] I. Olmez, N. K. Aras, G. E. Gordon and W. H. Zoller, *Anal. Chem.*, 1974, **46**, 1935.
[96] C. Block, R. Dams and J. Hoste, *Proc. Symp., IAEA Vienna, 15–19 March 1976*, p. 101.
[97] W. B. Renfro and W. A. Jester, *Trans. Am. Nucl. Soc.*, 1972, **15**, 718.
[98] J. A. Cooper, L. A. Rancitelli and T. M. Tanner, in *Pacific Northwest Lab. Annual Rept. 1972*, USAEC, Div. Biomed. Environ. Res., Vol. 2, Pt. 1, BNWL-1751, April 1973, pp. 62–63 and 112–113.
[99] J. Bogen, *Symp. on Nucl. Tech. in Comparative Studies of Food and Environmental Contamination*, Otaniemi, Finland, 27 August 1973, p. 75. IAEA, Vienna, 1974.
[100] J. Bogen, *Atm. Environ.*, 1973, **7**, 1117.
[101] B. Sansoni, A. Alian, R. K. Iyer and H. Petri, *Trans. Am. Nucl. Soc.*, 1979, **32**, 134.
[102] P. A. Cawse, *UK At. Energy Res. Estab. Rept.*, 1981, AERE-R 9886.
[103] W. Maenhaut, W. H. Zoller, R. A. Duce and G. L. Hoffman, *J. Geophys. Res.*, 1979, **84**, 2421.
[104] L. Bate, S. E. Lindberg and A. W. Andrew, *Trans. Am. Nucl. Soc. Suppl.*, 1975, **21**, No. 3, 20.
[105] L. Bate, S. E. Lindberg and A. W. Andrew, *J. Radioanal. Chem.*, 1976, **32**, 125.
[106] G. F. Clemente, *J. Radioanal. Chem.*, 1976, **32**, 25.
[107] R. H. Hammerle, R. H. Marsh, K. Rengan, R. D. Giauque and J. M. Jaklevic, *Anal. Chem.*, 1973, **45**, 1939.
[108] W. John, R. Kaifer, K. Rahn and J. Weselowski, *Atm. Environ.*, 1973, **7**, 1.
[109] J. R. Keane and E. M. R. Fischer, *Atm. Environ.*, 1968, **2**, 603.
[110] J. Lee and R. E. Jervis, *Trans. Am. Nucl. Soc.*, 1968, **11**, 1, 50.
[111] I. Obrusnik, B. Stárková and J. Blažek, *J. Radioanal. Chem.*, 1976, **31**, 495.
[112] N. J. Roberts, *Thesis*, UCRL-51499, 8 February 1974.
[113] F. C. Adams, M. J. van Crean, P. J. M. van Espen and D. Andreuzzi, *Atm. Environ.*, 1980, **14**, 879.
[114] J. M. Oudov, R. C. Ragaini and A. H. Biermann, *Environ. Sci. Tech.*, 1979, **13**, 946.
[115] J. Santroch and I. Obrusnik, *Radioisotopy*, 1973, **14**, 499.
[116] H. Vogg and R. Härtel, *Z. Anal. Chem.*, 1973, **267**, 257.
[117] H. Vogg and R. Härtel, *J. Radioanal. Chem.*, 1977, **37**, 857.
[118] S. S. Brar, *Intern. Conf. Modern Trends in Activation Analysis*, Vol. 1, 1968, p. 43.
[119] S. S. Brar, *Trans. Am. Nucl. Soc.*, 1969, **12**, 485.
[120] S. S. Brar, D. M. Nelson, J. R. Kline and P. F. Gustafson, *J. Geophys. Res.*, 1970, **75**, 2939.
[121] C. M. Gordon and W. H. Zoller, *Trans. Am. Nucl. Soc.*, 1973, **16**, 42.
[122] D. H. Klein, *Environ. Sci. Technol.*, 1975, **9**, 1973.
[123] E. T. Kucera and R. R. Heinrich, *J. Radioanal. Chem.*, 1976, **32**, 137.
[124] R. C. Ragaini and J. M. Ondov, *J. Radioanal. Chem.*, 1977, **37**, 679.
[125] F. K. Lepple, P. E. Wilkniss, D. J. Bressan, W. H. Cullen, G. W. Phillips and K. W. Marlow, *J. Radioanal. Chem.*, 1977, **37**, 661.
[126] W. Maenhaut and W. H. Zoller, *J. Radioanal. Chem.*, 1977, **37**, 637.
[127] T. Mamuro, Y. Matsuda and A. Mizohata, *Ann. Rep. Radiation Centre, Osaka Prefect.*, 1973, **14**, 7.
[128] T. W. Oakes, A. K. Furr, D. J. Adair and T. F. Parkinson, *J. Radioanal. Chem.*, 1977, **37**, 881.
[129] K. K. S. Pillay and C. C. Thomas, Jr., *J. Radioanal. Chem.*, 1971, **7**, 107.
[130] C. H. Westcott, *At. Energy Canada Ltd., AECL-1101*, 1960.
[131] P. Buat-Menard, C. E. Lambert, M. Arnold and R. Chesselet, *J. Radioanal. Chem.*, 1980, **55**, 445.
[132] E. Havránek, A. Bumbálová, M. Harangozó, E. Dejmková and F. Krečmer, *Chem. Zvesti*, 1983, **37**, 201.
[133] K. Masumoto and M. Yagi, *J. Radioanal. Chem.*, 1983, **78**, 233.
[134] A. R. Byrne and L. Kosta, *J. Radioanal. Chem.*, 1978, **44**, 247.
[135] D. Hoede, J. Zonderhuis and H. A. Das, *J. Radioanal. Chem.*, 1980, **56**, 199.
[136] H. L. Rook, *J. Radioanal. Chem.*, 1977, **39**, 351.
[137] D. R. Williams and J. S. Hislop, *J. Radioanal. Chem.*, 1977, **39**, 359.
[138] E. Häsänen and S. Salmela, *Radiochem. Radioanal. Lett.*, 1979, **37**, 207.
[139] I. Valente, M. J. Minski and H. J. M. Bowen, *J. Radioanal. Chem.*, 1978, **45**, 417.
[140] W. Zmijewska, *J. Radioanal. Chem.*, 1980, **58**, 367.
[141] A. Faanhoff and H. A. Das, *Radiochem. Radioanal. Lett.*, 1978, **33**, 403.

[142] M. Dermelj, J. Novak, V. Ravnik and L. Kosta, *J. Radioanal. Chem.*, 1978, **44**, 271.
[143] J. G. Farmer and J. D. Cross, *Radiochem. Radioanal. Lett.*, 1979, **39**, 429.
[144] A. A. Prapuolenis and J. M. Bakes, *Radiochem. Radioanal. Lett.*, 1969, **1**, 19.
[145] N. G. Zyrin, E. F. Kovnatskii, N. P. Roslyakov *et al.*, in *Yaderno-fizicheskie metody analiza v kontrole okruzhayushchei sredy*, p. 228. Gidrometeorizdat, Leningrad, 1985.
[146] M. J. Minski, C. A. Girling and P. J. Peterson, *Radiochem. Radioanal. Lett.*, 1977, **30**, 179.
[147] R. Stella, N. Genova and M. D. Casa, *Radiochem. Radioanal. Lett.*, 1977, **30**, 65.
[148] V. Kobayashi and K. Kudo, *J. Radioanal. Chem.*, 1978, **46**, 265.
[149] V. Kobayashi and K. Kudo, *J. Radioanal. Chem.*, 1979, **54**, 49.
[150] M. T. G. Valentini, N. Genova and R. Stella, *Radiochem. Radioanal. Lett.*, 1980, **44**, 359.
[151] M. Yagi and K. Masumoto, *J. Radioanal. Chem.*, 1984, **83**, 319.
[152] A. R. Byrne, *Radiochem. Radioanal. Lett.*, 1979, **40**, 1.
[153] M. Yamashita and N. Suzuki, *J. Radioanal. Chem.*, 1980, **60**, 73.
[154] T. C. Pung, C. C. Chou, H. T. Tsai and S. C. Wu, *J. Radioanal. Chem.*, 1981, **63**, 253.
[155] A. Faanhoff, H. A. Das and J. Zonderhuis, *J. Radioanal. Chem.*, 1980, **56**, 173.
[156] M. Kasrai, M. J. Shoushtarian and M. H. Bozorgzadeh, *J. Radioanal. Chem.*, 1977, **41**, 73.
[157] S. Ahmad, M. S. Chaudhary, A. Mannan and I. H. Qureshi, *J. Radioanal. Chem.*, 1983, **78**, 375.
[158] A. Faanhoff, and H. A. Das, *Radiochem. Radioanal. Lett.*, 1979, **41**, 367.
[159] T. Sato, T. Kato and N. Suzuki, *J. Radioanal. Chem.*, 1977, **36**, 221.
[160] U. C. Mishra and G. N. Shaikh, *J. Radioanal. Chem.*, 1983, **78**, 385.
[161] S. Ahmad, M. S. Chaudhary, and I. H. Qureshi, *J. Radioanal. Chem.*, 1979, **54**, 331.
[162] A. Faanhoff and H. A. Das, *J. Radioanal. Chem.*, 1980, **56**, 131.
[163] M. V. Sherif, R. M. Awadallah and A. E. Mohamed, *J. Radioanal. Chem.*, 1979, **53**, 145.
[164] G. D. Kanias and S. M. Philianos, *J. Radioanal. Chem.*, 1979, **52**, 389.
[165] Z. Abedinzadeh, A. Owlya and H. Rafii, *Radiochem. Radioanal. Lett.*, 1978, **32**, 315.
[166] P. Ila and P. Jagam, *J. Radioanal. Chem.*, 1980, **57**, 205.
[167] J. Kučera and J. J. M. DeGoeij, *J. Radioanal. Chem.*, 1981, **63**, 23.
[168] T. E. Gills, M. Gallorini, H. L. Rook, *J. Radioanal. Chem.*, 1978, **46**, 21.
[169] M. Ördögh, *J. Radioanal. Chem.*, 1978, **46**, 27.
[170] J. Kučera, *Radiochem. Radioanal. Lett.*, 1979, **38**, 229.
[171] R. A. Nadkarni, *Radiochem. Radioanal. Lett.*, 1977, **30**, 329.
[172] G. D. Vanias and S. M. Philianos, *J. Radioanal. Chem.*, 1978, **46**, 87.
[173] R. A. Nadkarni, G. H. Morrison, *Anal. Chim. Acta*, 1978, **99**, 133.
[174] N. B. Shtangeeva, V. D. Kulikov, L. V. Micheeva *et al.*, in *Yaderno-fizicheskie metody analiza v kontrole okruzhayushchei sredy*, p. 218. Gidrometeorizdat, Leningrad, 1985.
[175] Ch. Berthelot, G. Carraro and V. Verdingh, *J. Radioanal. Chem.*, 1980, **60**, 443.
[176] A. I. Ryabinin, L. V. Saltykova, I. I. Martemyanov *et al.*, in *Yaderno-fizicheskie metody analiza v kontrole okruzhayushchei sredy*, p. 97. Gidrometeorizdat, Leningrad, 1980.
[177] T. Kato and M. Kato, *J. Radioanal. Chem.*, 1980, **57**, 105.
[178] D. K. Teherani, H. Altmann, M. J. Michailov, M. S. Lazarova and D. A. Entschev, *J. Radioanal. Nucl. Chem. Lett.*, 1984, **85**, 99.
[179] 4N. van der Klugt, P. Poelstra and E. Zwemmer, *J. Radioanal. Chem.*, 1977, **35**, 109.
[180] H. U. Fusban, Ch. Segebade and B. J. Schmitt, *J. Radioanal. Chem.*, 1981, **67**, 101.
[181] P. S. Tjioe, J. J. M. Goeij and J. P. W. Houtman, *J. Radioanal. Chem.*, 1973, **16**, 153.
[182] G. H. Morrison and N. H. Potter, *Anal. Chem.*, 1972, **44**, 839.
[183] A. C. D. Leslie and H. Smith, *J. Radioanal. Chem.*, 1978, **42**, 17.
[184] W. Zmijewska and T. Semkow, *J. Radioanal. Chem.*, 1978, **46**, 73.
[185] M. Allegrini, R. Delfanti, M. Di Casa and E. Orvini, *Radiochem. Radioanal. Lett.*, 1983, **59**, 163.
[186] S. Ohno, *Analyst*, 1980, **105**, 246.
[187] J. J. Fardy and G. D. McOrist, *J. Radioanal. Nucl. Chem. Lett.*, 1984, **87**, 239.
[188] L.-Y. Sun, M. R. Conrady and R. A. Schmitt, *J. Radioanal. Nucl. Chem.*, Articles, 1984, **81**, 323.
[189] A. J. Blotcky, C. Falcone, V. A. Medina and E. P. Raek, *Anal. Chem.*, 1979, **51**, 178.
[190] R. Cornelis, J. Versieck, L. Mees, J. Hoste and F. Barbier, *J. Radioanal. Chem.*, 1980, **55**, 35.
[191] E. Sabbioni, E. Marafante, R. Pietra, L. Goetz, F. Girardi and E. Orvini, *Proc. Symp. Nuclear Activation Techniques in the Life Sciences*, p. 179, IAEA, Vienna, 1978.

[192] P. K. Shah and B. C. Haldar, *Radiochem. Radioanal. Lett.*, 1979, **39**, 301.
[193] R. Ganesan, A. K. Ghose and B. C. Haldar, *Radiochem. Radioanal. Lett.*, 1978, **35**, 153.
[194] S. Landsberger and E. Hoffman, *J. Radioanal. Nucl. Chem. Lett.*, 1984, **87**, 41.
[195] R. Raie and H. Smith, *J. Radioanal. Chem.*, 1979, **48**, 185.
[196] D. M. McKowan and J. S. Morris, *J. Radioanal. Chem.*, 1978, **43**, 411.
[197] L. W. Nicholson and K. Rengan, *J. Radioanal. Chem.*, 1979, **54**, 355.
[198] E. Orvini and R. Delfanti, *Radiochem. Radioanal. Lett.*, 1979, **37**, 199.
[199] Y. Minagawa and K. Kamegaya, *Radiochem. Radioanal. Lett.*, 1969, **1**, 69.
[200] J. D. Cross, I. M. Dale, A. C. D. Leslie and H. Smith, *J. Radioanal. Chem.*, 1979, **48**, 197.
[201] J. P. W. Houtman, M. De Bruin, J. J. M. de Goeij and P. S. Tjioe, *Proc. Symp. Nuclear Activation Techniques in the Life Sciences*, p. 599. IAEA, Vienna, 1978.
[202] R. E. Jervis and B. Tiefenbach, *Proc. Symp. Nuclear Activation Techniques in the Life Sciences*, p. 627. IAEA, Vienna, 1978.
[203] T. Sato, *J. Radioanal. Nucl. Chem. Lett.*, 1984, **86**, 141.
[204] O. Stulzaft, B. Maziere and S. Ly, *J. Radioanal. Chem.*, 1980, **55**, 291.
[205] S. S. Krishnan, E. Cortés, V. Cassorla, L. Muñoz and N. Gras, *J. Radioanal. Nucl. Chem. Lett.*, 1985, **93**, 327.
[206] J. N. Biso, J. M. Cohen and S. M. Resnizky, *Radiochem. Radioanal. Lett.*, 1983, **58**, 175.
[207] M. Kregar, J. Müller, P. Rupnik, V. Ramšak and F. Špiler, *Proc. Symp. Nuclear Activation Techniques in the Life Sciences*, p. 407. IAEA, Vienna, 1978.
[208] C. J. Evans, P. Cummins, J. Dutton, W. D. Morgan, A. Sivyer and R. R. Ghose, *Proc. Symp. Nuclear Activation Techniques in the Life Sciences*, p. 719. IAEA, Vienna, 1978.
[209] K. J. Ellis, D. Vartsky and S. H. Cohn, *Proc. Symp. Nuclear Activation Techniques in the Life Sciences*, p. 733. IAEA, Vienna, 1978.
[210] E. Cortés, V. Cassorla, L. Muñoz, N. Gras and S. S. Krishnan, *J. Radioanal. Nucl. Chem. Lett.*, 1981, **50**, 177.
[211] T. Izak-biran and V. P. Guinn, *J. Radioanal. Chem.*, 1980, **55**, 61.
[212] R. Zeisler and R. R. Greenberg, *J. Radioanal. Chem.*, 1982, **75**, 27.
[213] Z. B. Alfassi and N. Lavi, *Radiochem. Radioanal. Lett.*, 1982, **53**, 173.
[214] C. A. Weers, D. Hoede and H. A. Das, *J. Radioanal. Chem.*, 1978, **42**, 113.
[215] S. J. S. Anand, *J. Radioanal. Chem.*, 1978, **44**, 101.
[216] S. J. S. Anand, *J. Radioanal. Chem.*, 1978, **47**, 93.
[217] A. Brandone, P. A. Borroni and N. Genova, *Radiochem. Radioanal. Lett.*, 1983, **57**, 83.
[218] R. J. Rosenberg, *J. Radioanal. Chem.*, 1979, **50**, 109.
[219] A. R. Byrne, *Radiochem. Radioanal. Lett.*, 1982, **52**, 99.
[220] I. Horkay, D. K. Teherani, H. Altmann and Á. Kósa, *J. Radioanal. Nucl. Chem. Lett.*, 1985, **93**, 81.
[221] Z. R. Turel and B. C. Haldar, *J. Radioanal. Nucl. Chem.*, Articles, 1984, **84**, 109.
[222] H. S. Dang, H. B. Desai, D. D. Jaiswal, S. R. Kayasth and S. Somasundaram, *J. Radioanal. Chem.*, 1983, **77**, 65.
[223] R. O. Korob, I. M. Cohen, M. Loge and M. I. Milá, *J. Radioanal. Chem.*, 1978, **42**, 121.
[224] E. Cortés, N. Gras, L. Muñoz and V. Cassorla, *J. Radioanal. Chem.*, 1982, **69**, 401.
[225] O. Suschny, R. Dybczyński and A. Veglia, *J. Radioanal. Chem.*, 1982, **69**, 147.
[226] A. P. Grimanis, M. Vassilaki-Grimani, D. Alexiou and C. Papadatos, *Proc. Symp. Nuclear Activation Techniques in the Life Sciences*, p. 241. IAEA, Vienna, 1978.
[227] A. R. Byrne, L. Kosta, V. Ravnik, J. Štupar and V. Hudnik, *Proc. Symp. Nuclear Activation Techniques in the Life Sciences*, p. 255. IAEA, Vienna, 1978.
[228] C. Bergerioux and L. Zikovsky, *J. Radioanal. Chem.*, 1978, **47**, 173.
[229] D. J. Huljev, *J. Radioanal. Chem.*, 1982, **74**, 249.
[230] R. E. Williams, P. K. Hopke and R. A. Meyer, *J. Radioanal. Chem.*, 1981, **63**, 187.
[231] T. E. Gills and L. T. McClendon, *J. Radioanal. Chem.*, 1977, **39**, 285.
[232] A. Chattopadhyay, K. M. Ellis and Desilva K. Nimalasiri, *Proc. Symp. Nuclear Activation Techniques in the Life Sciences*, p. 667. IAEA, Vienna, 1978.
[233] T. Sato and T. Kato, *Radiochem. Radioanal. Lett.*, 1980, **42**, 227.
[234] L. De Reu, R. Cornelis, J. Hoste and S. Ringoir, *Radiochem. Radioanal. Lett.*, 1979, **40**, 51.
[235] D. Brune, G. F. Nordberg, P. O. Wester and B. Bivered, *Proc. Symp. Nuclear Activation Techniques in the Life Sciences*, p. 643. IAEA, Vienna, 1978.
[236] D. Behne and H. Jürgensen, *J. Radioanal. Chem.*, 1978, **42**, 447.
[237] H. Nakahara, Y. Nagame, Y. Yoshizawa, H. Oda, S. Gotoh and Y. Murakami, *J. Radioanal. Chem.*, 1979, **54**, 183.

[238] K. Kostadinov and R. Djingova, *J. Radioanal. Chem.*, 1981, **63**, 5.
[239] N. I. Ward and D. E. Ryan, *Anal. Chim. Acta*, 1979, **105**, 185.
[240] J. S. Hislop and D. R. Williams, *J. Radioanal. Chem.*, 1973, **16**, 329.
[241] L. M. Mosulishvili, V. Yu. Dundua, N. E. Kharabadze, E. Yu. Efremova and N. V. Chikhladze, *J. Radioanal. Nucl. Chem., Articles*, 1984, **83**, 13.
[242] M. M. Korobenkova, L. L. Pelekis and I. E. Tsirkunova, in *Yaderno-fizicheskie metody analiza v kontrole okruzhayushchei sredy*, p. 188. Gidrometeorizdat, Leningrad, 1980.
[243] M. Terai, A. Akabane, K. Ohno, S. Sakurai, F. Tsunoda, K. Hashimoto and G. Nishida, *J. Radioanal. Chem.*, 1979, **52**, 143.
[244] I. Obrusnik, B. Stárková, J. Blažek and V. Bencko, *J. Radioanal. Chem.*, 1979, **54**, 311.
[245] V. Galatanu and Ch. Engelmann, *J. Radioanal. Chem.*, 1982, **74**, 161.
[246] U. Tomza and W. Maenhaut, *J. Radioanal. Nucl. Chem. Lett.*, 1984, **86**, 209.
[247] S. Ohmori, H. Tsuji, Y. Kusaka, T. Takeuchi, T. Hayashi, J. Takada, M. Koyama, H. Kozuka, M. Shinogi *et al.*, *J. Radioanal. Chem.*, 1981, **63**, 269.

8

Non-activation interaction analysis

Non-activation interaction analytical methods are based on utilization of nuclear radiation (in some cases also radiation of non-nuclear origin) when the sample contains no radionuclides. They involve the interaction of nuclear radiation or X-rays with the sample, leading to absorption and back-scattering but not to activation of the analyte. Included are methods based on:

— elastic scattering of charged particles;
— absorption and back-scattering of beta-radiation;
— absorption and back-scattering of photon radiation;
— measurement of the intensity of X-ray fluorescence excited by primary radiation from radionuclides and charged particles;
— absorption and thermalization of neutrons;
— ionization of gases by nuclear radiation.

Most analytical methods based on the interaction of nuclear radiation with a sample are non-destructive and this is a significant advantage. Experimental and functional simplicity, as well as the speed of analysis, are additional advantages [1, 2].

From the point of view of environmental analysis, the most useful non-activation interaction techniques are X-ray fluorescence by photon or charged-particle excitation, ionization of gases by nuclear radiation, elastic scattering of charged particles and back-scattering of beta-radiation [3–5].

8.1 ANALYSIS BY ELASTIC SCATTERING OF CHARGED PARTICLES

Elastic scattering is a valuable tool for studying a variety of determinations in solid materials, e.g. surface layer composition, trace heavy metal concentrations and depth profile analysis.

In a charged-particle back-scattering measurement, the yield and energy of the scattered particles are directly related to the composition of the sample. This method is attractive because of its non-destructive character, its selectivity and the relative simplicity of the equipment used.

Analysis by elastic scattering of charged particles is based on a change in the ratio of the kinetic energy of the charged particle after scattering to that preceding scattering (which depends on the target nucleus mass number). The shape of the spectrum of back-scattered particles is affected by the thickness of the scattering material. When a thin target is bombarded with a beam of charged particles, a detector will record different particle groups originating from nuclear interactions. The spectrum of back-scattered particles is a succession of peaks of different amplitudes (Fig. 8.1a), each

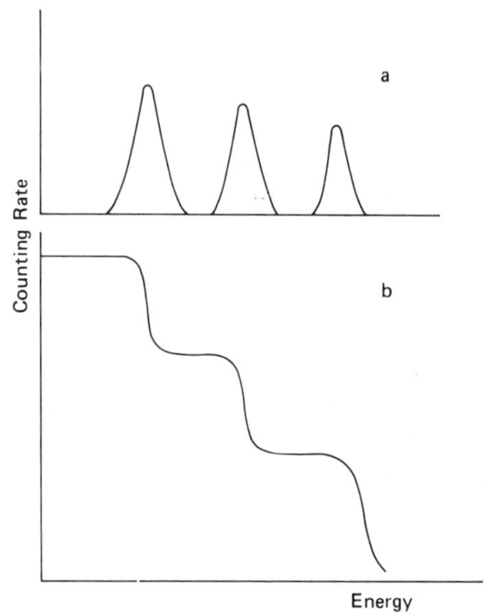

Fig. 8.1 — Hypothetical spectrum of back-scattered charged particles from a very thin (a) and a 'thick' (b) analysed sample.

peak corresponding to a given nuclear mass. In the study of thin films, foils, and surfaces, and in the analysis of solid compounds, scattered charged particles are detected at large angles ($>160°$). By surface, we mean the first layer, of less than 1 μm thickness. For a sample that is thick enough to be analysed, the spectrum of back-scattered particles is characterized by rough edges (Fig. 8.1b). The locations of the peaks or edges correspond to the elements present in the sample, and their heights depend on the amount of these elements present. Quantitative analytical evaluation of the spectrum may be done by using standards or appropriate calculations.

A simplified experimental arrangement for back-scattering measurement is shown in Fig. 8.2. It consists of a small scattering chamber with a sample holder electrically insulated for current measurement, a particle spectrometer with a linear energy response, and a multichannel analyser. Particles from protons up to $^{40}Ar^+$ and heavier ions have been employed in the past few years, but most often $^4He^+$ particles of moderate energy (0.5–3 MeV) or alpha-particles from radionuclide sources ($\simeq 5$ MeV) are used. The collimated particle beam impinges on the sample surface and the scattered particles are detected in a solid-state detector (surface barrier detector).

For heavy element detection, the alpha-particle back-scattering technique is preferred when the matrix is composed of light atoms, which is often the case for biological samples [6]. Elastic scattering of charged particles has been used for the analysis of aerosols collected on a Millipore filter [6–8].

The determination of elements in complex systems can be based on comparison with the back-scattered radiation intensity obtained with pure elements, or with mixtures of known quantitivative composition containing the element of interest.

8.2 ANALYSIS BY ABSORPTION OF BETA-RADIATION

Beta-radiation intensity is reduced by inelastic collisions during passage through matter. The mass attenuation coefficient is directly proportional to the electron density of the absorbing material, which may be expressed by the ratio of the atomic number to the atomic weight, Z/A, of the element. For samples containing a number of elements, compounds, or mixtures, the mass attenuation coefficient depends solely on the electron density; its effective value, $(Z/A)_{\text{eff}}$, can be expressed by:

$$(Z/A)_{\text{eff}} = \sum_i c_i \frac{Z_i}{A_i}$$

where c_i is the weight fraction, Z_i the atomic number, and A_i the atomic weight of the ith element.

The surface density of the absorbing material may be expressed as a product of its density ρ, and thickness d. At a constant thickness, for a given beta-ray source and a given experimental arrangement, the intensity of the transmitted radiation I_t may be expressed by

$$I_t = f\left(\rho \sum_i c_i \frac{Z_i}{A_i}\right)$$

The ratio Z/A ranges from about 0.5 to 0.4 for elements with low and high atomic numbers, respectively, and is unity for hydrogen. Thus, hydrogen shows twice as great an absorption capacity as do other elements. When the sample contains only hydrogen and one other element, the hydrogen

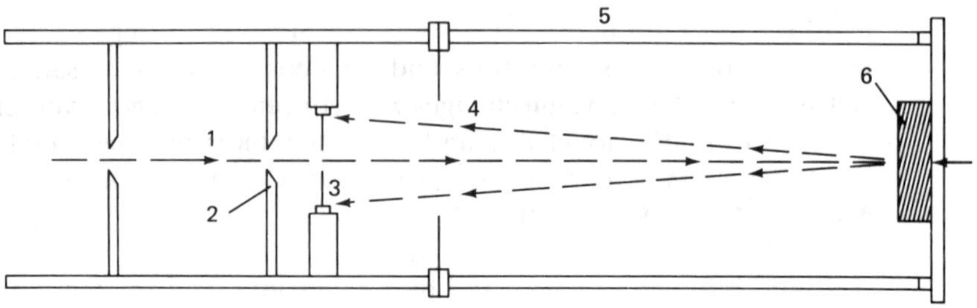

Fig. 8.2 — Simplified experimental arrangements for back-scattering measurements: 1 — projectile beam, 2 — collimator, 3 — annular detector, 4 — back-scattered particles, 5 — Faraday cup, 6 — sample, 7 — sample charge collection.

concentration can be determined with great precision by measuring the beta-radiation transmission for a known sample density.

Beta-ray absorption analysis of two-component systems can also be done on the basis of the change in sample density accompanying a change in chemical composition [1, 2]. A two-component system consisting of components with densities ρ_1 and ρ_2 has a density of

$$\rho = \rho_1 c_1 + \rho_2 c_2$$

where c is the weight fraction. Since densities ρ_1 and ρ_2 remain constant, the density of the sample will depend only on the concentration of its components. The greater the difference between the densities of the two components, the greater the sensitivity of the analysis. The thickness of the sample must be held constant during all measurements.

In beta-ray absorption analysis, the sample is situated between the radiation source (^{204}Tl, $^{90}Sr + ^{90}Y$, ^{85}Kr) and the detector. The radiation source is in a suitable holder, and the beta-particles emitted are collimated to hit the sample. With collimated beams, the source-to-detector distance has only a minor effect on the results, which is not the case with point sources.

The thickness of the sample layer analysed in the measuring cell depends on the maximum energy (E_{max}) of the beta-radiation employed. Each value of E_{max} corresponds to a certain optimum thickness of the sample which will provide maximum sensitivity.

Beta-ray absorption analysis has been used most successfully in the determination of hydrogen and the H:C ratio in liquid hydrocarbons. The percentage of hydrogen by weight and the H:C ratio are determined from beta-ray absorption and from the density, by making use of a calibration graph (plotted with the help of values obtained for pure hydrocarbons) or on the basis of an appropriate calculation.

On the basis of the change of $(Z/A)_{eff}$, the moisture content in various materials can be determined. In this way the moisture in various environmental samples has been determined.

The beta-ray absorption method has also been used to determine the concentrations of suspensions and aqueous solutions of salts. In these systems the $(Z/A)_{eff}$ value changes with concentration; however, changes in density are more noticeable and serve as a basis for these analyses. The sensitivity of these determinations is much poorer than that achieved in the determination of hydrogen [1, 2].

8.3 ANALYSIS BY BACK-SCATTERING OF BETA-RADIATION

A basic relationship in analysis by back-scattering of beta-radiation is represented by the dependence of the back-scattered beta-radiation intensity on the atomic number of the scattering material. When the scattering system consists of several elements, the intensity of the back-scattered radiation is a function of the effective (mean) atomic number of the scattering material (Z). This allows beta-back-scattering to be used for analytical purposes. Changes in the chemical composition are associated with changes in the effective atomic number, and thus with changes in the back-scattered radiation intensity.

The intensity of the back-scattered radiation increases strongly at first with increasing thickness, then the increase becomes more moderate, and after reaching a certain value, it remains unchanged, as the particles back-scattered in deeper layers are completely absorbed during their passage to the surface. The thickness above which further increase has no effect on the back-scattered radiation intensity, is termed the saturation thickness, d_s. In beta-back-scattering analysis a thickness of the material of interest which satisfies the condition $d > d_s$ is used, so that the effect of thickness of the scattering material is eliminated.

Present knowledge concerning beta-radiation back-scattering suggests that the beta-back-scattering method is applicable only to two-component systems. In a few cases it can also be used for analysing other samples, e.g. systems in which the atomic number (effective atomic number) of the analytes significantly exceeds the atomic number of the other components, so that all the remaining components can be considered together as the second component [1, 2, 9, 10].

In beta-back-scattering analysis, the intensity of the back-scattered radiation is measured under conditions which prevent the entrance of the original beta-rays into the detector. The basic types of geometrical arrangements of the system, radiation source (1), scattering material (2), and detector (3) are shown schematically in Fig. 8.3. A point source (^{204}Tl, $^{90}Sr + ^{90}Y$) of radiation is placed approximately on the vertical axis of the arrangement, between the scattering material and the detector (Fig. 8.3a). Alternatively, the scattering material is irradiated by collimated beams of beta-particles situated on one (Fig. 8.3) or both sides of it. Similar geometrical conditions can be obtained with a ring radiation source (Fig. 8.3c) surrounding the detector, the source being shielded by a grooved absorbing material. This arrangement produces a symmetrical radiation field.

The calibration graph method is used in beta-back-scattering analysis.

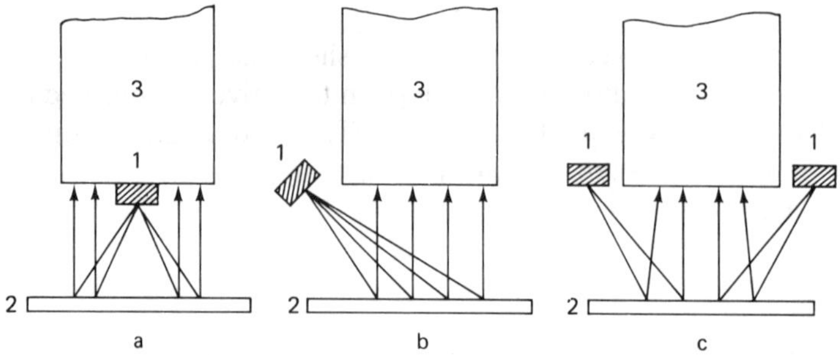

Fig. 8.3 — Basic types of the spatial arrangement of beta-back-scattering system: 1 — radiation source, 2 — scattering material, 3 — radiation detector; *a* — coaxial geometry, *b* — non-coaxial geometry, *c* — geometry with a ring-shaped radiation source.

The calibration graph is plotted as the dependence of the back-scattered radiation intensity on the concentration of one component. When a sample of thickness $d > d_s$, and the same radiation source and same geometry are used, the change of back-scattered radiation intensity can only be due to a change in the chemical composition of the scattering material.

Smal [11] used beta-back-scattering for environmental lead determination with the aid of a $^{90}Sr + ^{90}Y$ radiation source.

Determination of the ash content in coal is based on the assumption that the sample is a two-component system. There is a combustible component (represented by carbon) and several other elements (silicon, magnesium, aluminium, calcium and iron) which together form the ash content. Since the composition of the incombustible component remains essentially constant, it can be considered as the second component with a higher effective atomic number than carbon and can be determined by the beta-back-scattering method. The precision of the determination can be strongly affected by moisture, which should therefore be removed before measurements are made. The effect of grain size may be avoided by pulverizing the samples. An advantage of the back-scattering method for determining the ash content in coal, as compared to gamma-ray or X-ray absorption, is that the back-scattered beta-radiation intensity depends only on the effective atomic number of the sample and is essentially independent of sample density, which can markedly affect the results obtained by gamma-ray or X-ray absorption. The back-scattering method is simple and rapid.

The beta-back-scattering method has also been used to study the effect of pesticides on bird populations [12]. Ecological studies have shown a correlation between the decline in population of certain raptorial birds and their exposure to pesticides, possibly because of the thinner shells of the eggs of the exposed birds.

The eggshell-strength gauge [12] is based on the principle that the amount of beta-radiation back-scattered from an intact eggshell is a function of the shell strength. The gauge was laboratory tested [14, 15] to determine the effect of the base material (shell membrane, albumen and yolk) on the amount of radiation back-scattered by the shell, and the gauge readings were found to correlate well with shell thickness.

The laboratory data have been tentatively confirmed in a field test [16] with Japanese quail fed with DDT. The thickness of their eggs, measured with a micrometer, was less than in normal eggs, and the calcium content determined by absorptiometry was lower. Beta-back-scatter measurements of strength were then made and found to correlate well with both the thickness measurements and the calcium content determinations.

Similar tests were made [14] on the shells of eggs of Mallard ducks fed with DDE and DDD. In this case the only shells available for beta-back-scatter testing were those saved for residue analysis, but again thickness measurements correlated well with back-scatter counts.

8.4 ANALYSIS BY ABSORPTION AND BACK-SCATTERING OF PHOTONS

For utilization of the interaction of samples with photons emitted by radionuclides, two basic phenomena are used: absorption and back-scattering [1, 2].

Analysis based on absorption of photon radiation is recommended in two applications: the determination of hetero atoms in hydrocarbons, and in the determination of heavy elements. Absorption analysis assumes that the radiation penetrates the sample, and requires the sample to be accessible from both sides. For example, the determination of sulphur is based on the differences between the mass attenuation coefficients of sulphur, carbon, and hydrogen, and also on the fact that other elements are present in the sample only in minute quantities [18, 19].

Gamma-sources (^{170}Tm) and a bremsstrahlung source [(^{90}Sr + ^{90}Y)/U] have been used for lead determination [20]. GM and NaI(Tl) scintillation detectors were used for detection of the transmitted radiation.

The sensitivity of photon absorption analysis decreases with diminishing differences in the values of the mass attenuation coefficients for individual elements in the sample. On the other hand, when this difference increases, the concentration range in which the element can be determined is decreased.

For analysis by back-scattering of photon radiation, access is required from only one side of the sample. The method can be used for the determination of heavy elements, but this technique is only applicable with large samples and is not very sensitive [11].

Back-scattering of photon radiation is often used for the determination of ash content in coal [21]. The determination is based on differences in the atomic numbers of the elements forming the combustible and non-combustible components of coal. The effective atomic number of coal is about 6. The mineral content, owing to the effective atomic number being higher (~ 13) because of its Al and Si content, mainly aborbs gamma-radiation, and scattering is less effective. The opposite is true for carbon.

Absorption and back-scattering of photons is useful for rapid analysis, especially in routine analysis, in which the sensitivity requirements are not too great.

8.5 X-RAY FLUORESCENCE ANALYSIS

X-Ray fluorescence analysis is one of a number of X-ray spectrometric techniques available to determine the concentration of elements in different materials.

All types of X-ray spectrometry involve the bombardment of a sample with radiation of sufficient energy to eject inner-shell electrons from the atoms in the sample. The ejected electrons are immediately replaced by electrons from outer atomic shells, with simultaneous emission of X-rays with energies characteristic of the electron transitions (and thus of the elements) responsible for their emission. Variations of this basic process primarily lie in the nature of the bombarding radiation, and the method of detection and energy discrimination of the emitted X-rays.

Two types of X-ray detection systems are employed in X-ray spectrometry. The first, which has been used analytically for many years, is wavelength-dispersive detection, in which the X-rays from the specimen are highly collimated and dispersed according to the Bragg Law by a diffracting crystal before being measured at an angle corresponding to the characteristic wavelength. In the second type of detection (energy-dispersive spectrometry), X-rays from the specimen are incident on a detector that gives an output charge pulse precisely proportional to the energy of each X-ray photon absorbed.

The greatest variations among X-ray spectrometric methods are due to the different types of exciting radiation that can be employed [2].

X-Ray fluorescence may be induced by photon or charged-particle irradiation, usually with protons. Thus we may distinguish between X-ray fluorescence (excitation by photons) — XRF and particle induced X-ray emission (usually induced by protons) — PIXE. Figure 8.4 gives a schematic representation of these two possibilities [4, 21]. The cross-sections for the excitations are of the same order of magnitude. The beam intensity however, differs considerably. In photon excitation it is $\sim 10^7$/sec, whereas proton beams of up to 5×10^{11}/sec are feasible. The limiting factor is the cooling of the target [4].

The sensitivity of PIXE is much higher than that of XRF. In both techniques there is an appreciable background; in XRF it is due to scattering, in PIXE to bremsstrahlung caused by the slowing down of the incident protons [4, 22].

Excitation by X-ray tubes or by radiation from radionuclides is used for X-ray fluorescence elemental analysis.

The simplest method using an X-ray tube is the direct mode with its bremsstrahlung spectrum for the excitation of characteristic X-rays from the sample. Conventional X-ray spectrometers at present range from simple, manually-operated instruments to fully automatic equipment capable of analysing a hundred or more samples without human intervention.

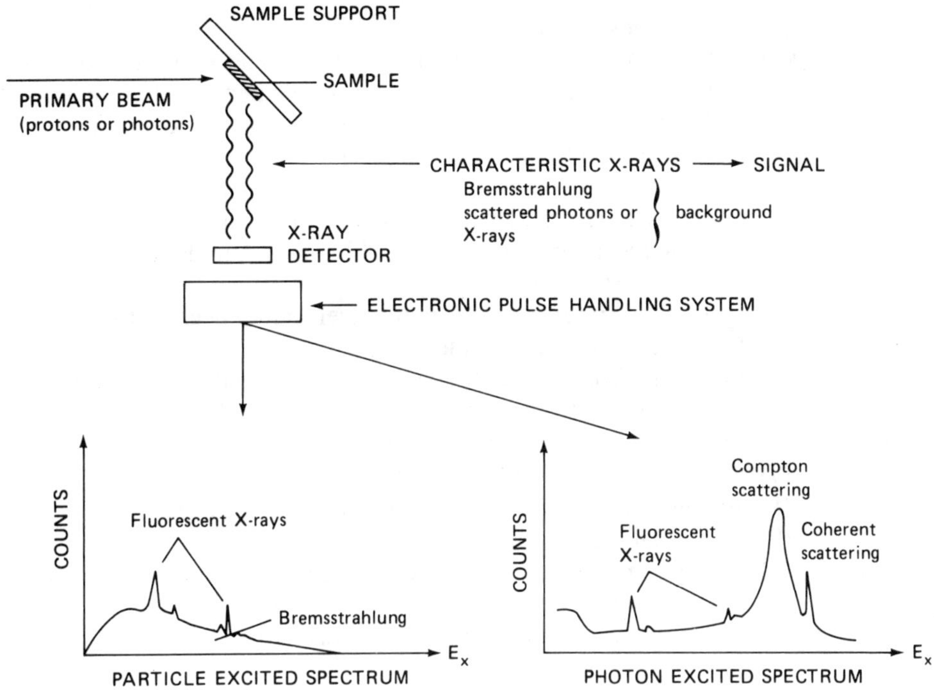

Fig. 8.4 — Schematic representation of the major components of an energy-dispersive X-ray fluorescence analysis system.

Instead of X-ray tubes, radiation from radionuclides is often used for excitation of fluorescent X-rays. Radionuclide X-ray fluorescence analysis has a number of advantages over that with X-ray tubes:

— simple radiation sources;
— wide energy of monochromatic photon radiation (X-ray tubes cannot cover the same energy range of excitation radiation as radionuclides);
— constant intensity of radionuclide radiation,
— simple experimental arrangement and easy operation,
— commercial availability of equipment.

Conventional X-ray fluorescence analysis with X-ray tubes is characterized by high sensitivity, but expensive and complicated apparatus is required. For many practical analytical tasks extremely high sensitivity is not necessary, and instead of X-ray tubes radionuclide sources of radiation may be used.

Radionuclide X-ray fluorescence analysis provides a potentially useful technique for the multielement analysis of environmental samples. It has some advantages over the other analytical methods presently being used by researchers in the environmental sciences:

(1) it is non-destructive;
(2) little or no chemical treatment of sample is needed prior to analysis;
(3) nearly simultaneous determination of elements with atomic number >20 with very high potential sensitivity (dependent upon further investigation into instrumental design);
(4) for biological samples, matrix interference effects such as self-absorption and enhancement of X-ray radiation are comparatively small because the major constituents of biological samples are light elements such as C, H and N;
(5) spectral line interferences are relatively minor because there are only a few X-ray lines for each element;
(6) the XRF spectrum of an element, (except that obtained with a high-resolution spectrometer), is generally independent of the physical state and chemical form of the element;
(7) analysis time ranging from 1 sec to several minutes;
(8) simplicity of analytical operation and data retrieval;
(9) no specialized facilities are required;
(10) the equipment is potentially portable (when solid-state detectors which do not need to be cooled to liquid-nitrogen temperature become available, or when other suitable detectors — scintillation or proportional — are used);
(11) the method can easily be automated;
(12) relatively low costs for instrumentation, maintenance and time invested per sample analysed.

A rather simple experimental arrangement can be used for radionuclide X-ray fluorescence, as shown in Fig. 8.5 [23]. The nuclide source emits high-

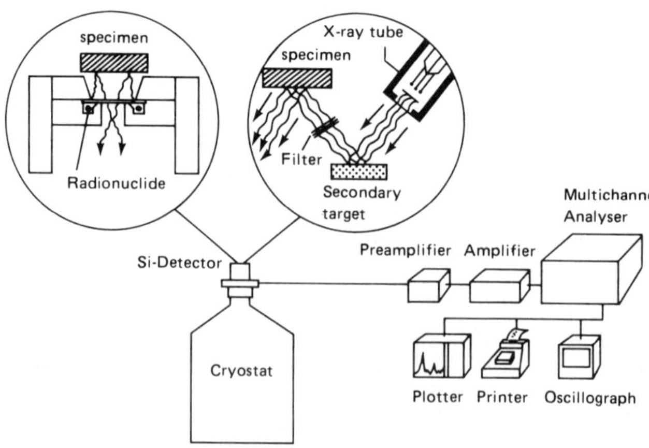

Fig. 8.5 — Energy-dispersive X-ray fluorescence. Left, excitation by radionuclides; right, excitation by X-ray tubes in combination with secondary targets. (Reproduced from [23] by permission of Professor K. H. Lieser).

energy radiation, which bombards the sample. Fluorescent X-rays from the sample are analysed with a silicon detector and multichannel analyser. Samples may be compacted or in powder form, and matrix corrections are usually made by means of the Compton scattering peak. Calibration standards for environmental samples are sometimes prepared from silica gel (for silicates or soils) or flour (for plant materials).

Many radionuclides can be used for X-ray excitation, and a partial list is given in Table 8.1 [23]. The most commonly used nuclides are ^{109}Cd and

Table 8.1 — Radionuclide sources for X-ray fluorescence analysis (Reproduced from [23] by permission of Professor K. H. Lieser)

Radionuclide	Exciting radiation	Range of application (atomic number, Z)
^{3}H	β^{-}(18.4 keV)/Bremsstrahlung	12–35
^{55}Fe	Röntgen (5.95 keV)	9—30
^{57}Co	γ(122.1 + 136.4 keV)	70–94
^{71}Ge	Röntgen (9.2 + 10.2 keV)	22–29/50–64
^{75}Se	γ(121.1–400.7 keV)	66–94
^{85}Kr	γ(514.0 keV)	80–94
^{90}Sr/^{90}Y	β^{-}(500 + 2300 keV)/Bremsstrahlung	19–94
^{109}Cd	Röntgen (22.1; 25.0; 25.5 keV)	23–46/68–94
^{125}I	Röntgen (27.2–31.7 keV); γ(35.3 keV)	33–51/73–94
^{145}Pm	Röntgen (36.9–43.3 keV)	35–56/85–94
^{147}Pm(Al)	β^{-}(200 keV)/Bremsstrahlung	19–94
^{153}Gd	Röntgen (40.9–48.3 keV); γ(97.5 + 103.2 keV)	60–94
^{170}Tm	Röntgen (51.5–59.3 keV); γ(84.3 keV)	40–60/73–78 90–94
^{181}W	Röntgen (56.3–67.0 keV)	40–64/90–94
^{197}Hg	Röntgen (9.7; 67.0–80.2 keV); γ(191.4 + 268.7 keV)	20–27/46–74/ 84–94
^{210}Pb	Röntgen (10.8 keV); γ(46.5 keV)	20–29/46–66 32–58/73–94
^{210}Po	α(5.3 MeV)	12–17
^{238}Pu	Röntgen (13.6 keV)	23–38/56–82
^{241}Am	γ(59.5 keV)	28–69
^{244}Cm	Röntgen 14.3 keV	20–33/45–76

^{241}Am. Detection limits for various methods of excitation are given in Fig. 8.6 [23]. Detection limits vary between 1 and 10 ppm, and there are no great differences between X-ray tube and radionuclide XRF.

The recent progress of radionuclide X-ray fluorescence analysis shows that with radionuclide sources of radiation, proportional counters, solid-state detectors, and pulse-height analysers, fairly high sensitivity has been reached, comparable with that obtained with X-ray tubes [24].

Numerous publications demonstrate the usefulness of radionuclide X-ray fluorescence analysis for determining metal particulates in air. The analytical system often used consists of an annular radionuclide source (^{241}Am) a cryogenically-cooled Si(Li) detector and a small computer for

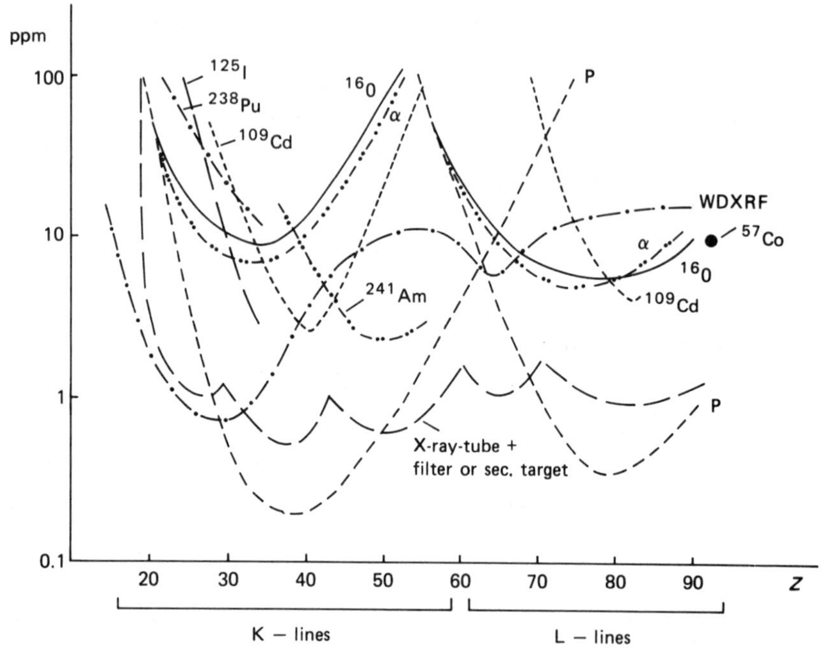

Fig. 8.6 — Detection limits of energy-dispersive X-ray fluorescence analysis with various methods of excitation. p,α, ^{16}O; excitation by an X-ray tube (various filters or secondary targets); excitation by radionuclides; ^{109}Cd, ^{241}Am, ^{125}I, ^{238}Pu, ^{57}Co and for wavelength-dispersive analysis. (Reproduced from [23] by permission of Professor K. H. Lieser).

collecting and processing spectra. Thin-film specimens are deposited on filter paper disks. The following elements have been determined: Ca, Ti, V, Cr, Mn, Fe, Co, Ni, Cu, Zn, Hg, Pb, As, Br, Sr, Zr and Mo [25].

Lead from automobile exhausts is a regular component not only of urban air, but also of soil and vegetation growing near a road with heavy traffic. When a proportional counter is used, the bromine fluorescence (bromine is always associated with lead from automobile exhausts) overlaps a portion of the lead fluorescence. By use of balanced filters, all the fluorescence due to bromine can be excluded, but so is part of the fluorescence due to lead. The result is a decreased lead sensitivity. For this reason many authors have used solid-state Ge(Li) or Si(Li) detectors. There are important advantages in using these detectors rather than the balanced filter method; however, the necessary cryogenic cooling of the detector is troublesome [26–33].

The process of welding is a significant source of pollution of working environments with heavy metals. Evaluation of the chemical quality of the working environment was made in our laboratory by means of radionuclide X-ray fluorescence analysis. Welding aerosols were sampled by a filtration method on Synpor 4 membrane ulterfilters. For excitation a ^{241}Am annular source was used. Fe, Mn, Cr, Ni, Co, V, Ti, Cu, Zn, Mo and W were determined in the welding aerosols which are produced in manual arc welding. These metals contribute directly to the problem of the welder's health protection and determine some economic aspects of the production process (ventilation equipment, etc.) [34, 35].

Analysis of water is another application of radionuclide X-ray fluorescence [36]. The use of this method for direct analysis of water is rare because of its somewhat limited sensitivity for bulk aqueous samples, which necessitates an enrichment step before the analysis. Nevertheless, it is applicable to water samples with relatively high concentrations of the elements to be determined, especially of one (or a few) predominant elements (waste waters, some natural waters). The advantage of this technique is the short time needed for the analysis. It is practically equal to the measuring time, as the preparation of the sample consists only of pouring the solution into a measuring cup and inserting it into the measuring head.

Many inorganic pollutants may be present in water in trace amounts. In such cases direct measurement is impossible, and before analysis, separation and preconcentration procedures are needed. For example, strong acid ion-exchange disks have been used for determination of trace metals in fresh water and sea-water. The *K*- and *L*-fluorescence radiation of elements present in the sample was excited by ^{109}Cd, ^{238}Pu and ^{241}Am radionuclide sources and counted by a proportional counter connected to a single-channel pulse-height analyser or by a solid-state Si(Li) detector connected to a multichannel analyser. Radionuclide X-ray fluorescence analysis has been used for the determination of trace elements in various types of water [37–52].

X-Ray fluorescence analysis is used for the simultaneous determination of metals in both the above-ground and underground parts of plants. Samples for analysis are prepared by dry ashing, wet ashing or by a non-destructive technique (pressing into pellets). For the determination of heavy elements ^{238}Pu, ^{109}Cd (determination of Mn, Fe, Ni, Cu, Zn, Pb) and ^{125}I (determination of Br, Sr) are most often used as excitation sources [27]. XRF has been used for the determination of trace elements in various plant materials [53–64].

In vivo X-ray fluorescence analysis has important applications in occupational and environmental medicine, e.g. for direct measurements of Pb and Cd contamination.

Lead poisoning remains a health hazard to children living in an urban environment. Excessive lead exposure results in permanent nuerological damage to the developing central nervous system of young children. The lead stored in teeth and bones is permanent and is related to the quantity of ingested lead. The lead can be assayed by measuring its characteristic X-ray radiation in a tooth irradiated *in situ* with gamma-rays from a ^{57}Co source. The sensitivity of the method is adequate to measure the lead levels commonly associated with the environmental background. The X-ray exposure of the tooth necessary for this determination is approximately one tenth of that required for a routine dental examination [67–72].

The X-ray fluorescence spectrum of hair, in view of the greater number of elements present, might be employed as a sort of "fingerprint" for police use. Perhaps more noteworthy is the usefulness of this method in the field of preventive and occupational medicine. For example, the analysis of hair samples to detect possible poisoning in urban areas presents obvious advantages regarding early recognition of individuals in a toxic state [73–76].

XRF is often used for the analysis of soil and sediments [73, 77–85].

X-Ray fluorescence analyses of all the important types of environmental materials (air, water, soil, sediments, biological materials, food), which are of interest to ecologists have been described [24].

Multielemental analysis of trace elements by particle-induced X-ray emission (PIXE) is useful for environmental research. PIXE is suitable for simultaneous determination of 10–15 elements present in amounts between 10^{-8} and 10^{-11} g in small samples which can be mounted on supporting foils (~ 5 mg/cm^2) in an area of the order of the cross-section of a uniform accelerator beam ($\lesssim 1$ cm^2).

In a typical analysis, a sample is irradiated for 1–10 minutes by protons of a few MeV energy from a van de Graaf accelerator, at beam currents of 1–500 nA. The spectrum recorded by an Si(Li) detector is stored for subsequent computer evaluation.

The low detection limits of PIXE depend on favourable signal-to-background ratio, the relative X-ray production efficiency for characteristic and continuum radiation, and the availability of intense fluxes of excitation radiation, e.g. focused low energy proton beams from van de Graaf accelerators [86].

PIXE is not a universally applicable analytical method but is very powerful for the analysis of aerosols [4, 86–94], plants [95–100] or biological materials [4, 86, 87, 101–113]. The elements which are determinable in biological materials range from Cl to Sr. Usually Mn, Fe, Cu, Zn, Br, Rb and Sr are assayed [88, 89].

The advantage of PIXE lies in its speed and the fact that only small aliquots are needed. Its limitations are the requirement for a thin and even distribution of the material over the support, and the small range of elements which can be determined.

8.6 ANALYSIS BY THERMALIZATION AND SCATTERING OF NEUTRONS

The neutron-scattering cross-section for all elements is almost constant in the epithermal region (1–300 eV). However, for light nuclei, a regular increase in scattering cross-section is found in this energy region (up to 2–6 barns). An important exception to this rule is hydrogen, the cross-section of which is 80 barns for epithermal neutrons. Hydrogen can also remove a considerable part of the energy of a neutron during a single collision. As a result, the greater the hydrogen content in the sample, the faster the moderation of neutrons. For example, the slowing-down power of water is many hundreds of times greater than that of oxygen, silicon, aluminium and other elements [1,2,114,115]. The method is convenient for moisture determination, for oil prospecting, or for determination of the C/H ratio in hydrocarbons.

The advantages of moisture determination by neutron methods are:

— the results are almost independent of the chemical composition when elements with large absorption cross-sections or larger amounts of hydrogen bound organically, or in other ways, are not present in the sample;
— the results are not influenced by external conditions such as temperature, pressure, or pH;
— the results are representative of the whole sample, since large volumes of material are analysed.

The neutron probe unit consists of a fast-neutron source and a detector of slow neutrons. In neutron moisture gauges, radionuclide neutron sources are used. Neutron probe or moisture gauges may be classified according to the detector used. Neutron moisture gauges working with activation detectors, scintillation counters, proportional BF_3-counters and highly sensitive ionization chambers have all been used. Activation detectors are seldom used but BF_3 proportional counters are used frequently.

Sub-surface, surface, and transmission neutron probes are all used. Sub-surface probes yield a fivefold higher counting rate, and hence a better statistical significance, than do surface probes. Sub-surface probes are used mainly for discontinuous measurement of loose materials. Surface probes allow continuous contactless measurements, and a rapid check of moisture in closed vessels. The transmission methods require that the material examined be accessible from both sides.

When moisture content is determined by means of neutron thermalization and scattering, it is the neutron flux density rather than the water content that is actually measured. It is a good idea to construct a calibration chart for each particular material, i.e., to plot the neutron flux density *vs*. the moisture per unit volume. Although any neutron probe can be used to make a general calibration curve for moisture content (in volume %), for accurate results it is desirable to calibrate under special operational conditions, because the readings of the neutron moisture gauge are influenced by the nature of the material, the dimensions of the measured volume, the wall thickness and the diameter of the access tube.

Neutron moisture gauges are used mostly for soil moisture determination. The precision of the neutron method is about the same as that of the gravimetric method and better than the precision of the tensometric and resistance methods. The neutron method has been used for moisture content determinations in live trees and agricultural products.

8.7 NEUTRON ABSORPTION ANALYSIS

Elements with large absorption cross-sections for thermal neutrons (e.g. Gd, B, Cd, Li, Hg, In, Cl) may be determined non-destructively by absorption of neutron radiation.

The methods of neutron absorption analysis can be classified according to various criteria. From the point of view of the evaluation of measurements, we have absolute and relative methods, whereas with respect to neutron energy, the methods are divided into those using polyenergetic (unfiltered) neutrons and those using monoenergetic neutrons.

Standard samples are used in the relative methods. The standards should have compositions similar to those of the samples. In the absolute method, the results are obtained from known values of absorption cross-sections and neutron fluxes, but this method is rather troublesome.

Samples containing only one element with a large absorption cross-section for slow neutrons can be analysed by measuring the absorption of the total neutron flux (polyenergetic neutrons). Neutron sources of low intensity may be used for this purpose, i.e., radionuclide neutron sources. In this way elements such as Li, B, Cd, Sm, Eu, Gd and Dy can be determined in environmental samples [1,115]. Since only minute amounts of radioactive nuclides are formed when Li, B, Cd and Gd are irradiated, the neutron absorption method is complementary to activation analysis.

The method using monoenergetic neutrons (of defined energy) is suitable for the determination of elements with a pronounced resonance absorption of neutrons (e.g. Ag, In, Rh, Ir, Au). The attenuation is measured only in the resonance energy region. The presence of other elements does not interfere, because the resonance absorption regions rarely overlap. In spite of these advantages, the methods are not extensively used, because a monoenergetic neutron flux can only be obtained with difficulty. The flux of monoenergetic neutrons can be monitored by mechanical selectors, selectors with a cyclotron-modulated beam and crystal monochromators.

8.8 ANALYSIS BASED ON THE IONIZATION CAUSED BY NUCLEAR RADIATION

The ionization of gases by nuclear radiation can be determined by conductivity measurement and may be used for the determination of various parameters of gases, mainly density and composition. Such measurements are used for the analysis of gaseous mixtures, e.g. to determine their density, pressure and flow-rate.

The analysed gas flows between two electrodes in the ionization chamber. The gas is ionized by radiation emitted from the source mounted in the chamber. The analytical signal is an ionization current passing through the chamber when a constant voltage is applied across the electrodes.

In the case of a binary gaseous mixture, the intensity of the ionization current, with constant voltage across the electrodes, is unambiguous; under certain conditions it is also a linear function of the concentration of one component. For this reason, ionization analysis is useful in gas chromatography, because the individual components flowing with the carrier gas have already been separated and may be detected quantitatively.

The ionization devices used in gas chromatography are referred to as *ionization detectors*; devices analysing mixtures of gases or vapours without the use of gas chromatography are called *ionization analysers* [1, 116, 117].

The basic scheme of an alpha gas-analyser is shown in Fig. 8.7. Two

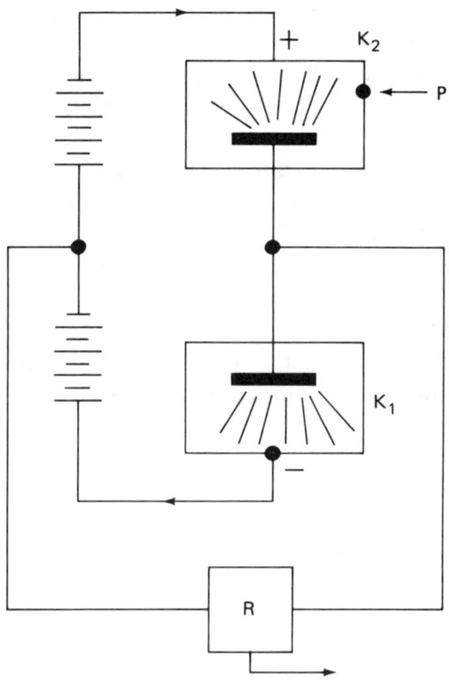

Fig. 8.7 — Schematic diagram of an alpha gas-analyser: (K_1 — chamber with standard gas, K_2 — flow-cell for analysed gas, P — inlet of gas, R — relay.

ionization chambers connected in parallel (difference mode) are used. The same amount of ^{210}Po or ^{232}Th is deposited on the electrode of both chambers. In one chamber, air or another convenient gas of standard composition is sealed, and the air or other gas sample flows through the second chamber. As long as the chemical composition of the gases in both chambers is the same, the ionization will also be the same and the galvanometer will show zero current. If a gas of some other composition enters the open chamber, the balance of ionization currents will be disturbed and the galvanometer will show a deflection. Since the ionization current passing through the chamber is proportional to the molecular weight of the gas, the galvanometer will measure and automatically register the concentration of the mixture, provided that the average molecular weight of the latter differs considerably from that of the reference gas.

In early applications of this method, minute concentrations of various gases in air were determined. With the reference chamber filled with air, less than 0.01% v/v of added gas (Cl_2, SO_2, NH_3, alcohol and vapours of fatty acids) may be determined.

Alpha ionization-analysers are used for the determination of harmful components in the atmosphere at working sites with personnel-safety hazards. With an alpha analyser the presence of CO, CO_2, SO_2, NH_3 and CH_4 in the atmosphere may be detected. The analysers can therefore be used successfully in chemical or coking plants, in mines, and for the detection of the infiltration of inflammable or harmful gases into the atmosphere. ^{226}Ra, ^{210}Po, ^{241}Am or ^{239}Pu are used as radiation sources [116].

A special class of ionization gas-analysers is that of radioactive fire-detectors. These are based on the fact that minute amounts of combustion products will depress the ionization current when entering the ionization chamber. This is explained by the size of the molecules of combustion products which, being greater than those of air, support the recombination of ions, which in turn causes the depression of the ionization current. The advantage of these detectors is that unlike most others, they can detect smouldering before flames appear [2, 118].

Beta-sources, e.g. ^{85}Kr and ^{63}Ni, are used in ionization analysers at a steadily growing rate.

Ionization by beta-radiation is utilized mainly in gas chromatography to detect and determine fractions in the carrier gas. The detectors work on the principle described for alpha ionization-analysers. The principles and functions of these detectors are described in detail in many publications on gas chromatography.

REFERENCES

[1] J. Tölgyessy and Š. Varga, *Nuclear Analytical Chemistry*, Vol. IV, *Analysis Based on the Interaction of Nuclear Radiation with Matter*, University Park Press, Baltimore; Veda Bratislava, 1975.

[2] J. Tölgyessy and M. Kyrš, *Radioanalytical Chemistry*, Veda, Bratislava; Ellis Horwood, Chichester, in the press.

[3] J. Tölgyessy, Yu. V. Yakovlev and G. H. Bilimovich, *Diagnostika okruzhayushchei sredy radioanaliticheskii metodami*, Energoatomizdat, Moscow, 1985.

[4] H. A. Das, A. Faanhof, and H. A. van der Sloot, *Environmental Radioanalysis*, Elsevier, Amsterdam, 1983.

[5] J. Tölgyessy, *Non-activation Interaction Analysis of Environmental Samples*, Consultants' Meeting on Advances in Nuclear Techniques for Analysis of Environmental Samples, Karlsruhe, 19–21 February 1986.

[6] G. Deconninck, *Introduction to Radioanalytical Physics*, Elsevier, Amsterdam, 1978.

[7] S. Rubin, *Anal. Chem.*, 1957, **29,** 736.

[8] M. A. Nicolet, J. W. Mayer and I. V. Mitchell, *Science*, 1972, **177,** 841.

[9] J. Lesný and J. Tölgyessy, *Chem. Listy*, 1975, **69,** 703.

[10] V. Kriváň, *Chem. Listy*, 1968, **62,** 129.

[11] Zb. Smal, *Pr. Cent. Inst. Ochrony Pracy*, 1971, **21,** 185.

[12] M. Gered, *Isotop. Radiat. Technol.*, 1971, **8,** 395.

[13] *Isotope Radiat. Technol.*, 1968, **5,** 284.

[14] P. W. Voisey and P. E. James, *Canad. Agr. Eng.*, 1970, **12,** 48.

[15] J. Bitman, H. C. Cecil, S. J. Harris and P. E. James, *Poultry Sci.*, 1969, **48,** 2184.

[16] J. Bitman, H. C. Cecil, S. J. Harris and G. F. Frieg, *Nature*, 1969, **224,** 44.

[17] R. G. Heath, J. W. Spann and S. F. Kreitzer, *Nature*, 1969, **224,** 47.

[18] A. P. Pyrah, R. S. Robertson and J. Woseman, *Anal. Chem.*, 1961, **13,** 1355.
[19] Saunders-Roe, Nuclear Enterprises Ltd., Publication No. S.P. 524, 525, England.
[20] T. Florkowski, *Postepy Techniky Jadrowej*, 1962, No. 4.
[21] S. Koch and W. Liebig, *Isotopentech.*, 1962, **2,** 353.
[22] W. Herda, *Nucl. Inst. Methods*, 1979, **158,** 587.
[23] K. H. Lieser, *Nuclear Analytical Techniques and their Application to Environmental Samples*, Consultants' Meeting on Advances in Nuclear Techniques for Analysis of Environmental Samples, Karlsruhe, 19–21 February 1986.
[24] J. Tölgyessy, E. Havránek and E. Dejmková, *Rádionuklidová röntgenfluorescenčná analýza zložiek životného prostredia (Radionuclide X-Ray Fluorescence Analysis of Environmental Components*), Alfa, Bratislava, 1983.
[25] E. Havránek, A. Bumbálová and E. Dejmková, *Chem. Zvesti*, 1980, **34,** 373.
[26] E. Havránek and E. Bumbálová, *Chem. Zvesti*, 1981, **35,** 57.
[27] E. Havránek, E. Dejmková and A. Bumbálová, *Acta Facult. Pharm. Univ. Commenianae*, 1983, **37,** 39.
[28] M. Gallorini, N. Genova, E. Orvini and R. Stella, *J. Radioanal. Chem.*, 1976, **34,** 135.
[29] M. Boazi, J. Gabay, S. Devir and C. Shenberg, *J. Radioanal. Chem.*, 1978, **45,** 395.
[30] T. Paradellis, *Nucl. Instr. Methods*, 1977, **140,** 205.
[31] H. R. Bowmann, J. G. Conway and F. Asaro, *Environ. Sci. Technol.*, 1972, **6,** 558.
[32] J. D. Bowmann and J. C. Vanderleyden, *Nucl. Instr. Methods*, 1970, **85,** 19.
[33] E. Dejmková and P. Schiller, *Acta Facult. Pharm. Univ. Commenianae*, 1979, **33,** 195.
[34] Ľ. Oláh and J. Tölgyessy, *J. Radioanal. Nucl. Chem. Lett.*, 1985, **93,** 43.
[35] J. Tölgyessy, Ľ. Oláh and E. H. Klehr, *Proc. Modern Trends in Activation Analysis*, 23–27 June 1986, Copenhagen, MTAA-7050.
[36] E. Havránek, M. Harangozó, M. Piatrik and J. Tölgyessy, *Radiochem. Radioanal. Lett.*, 1982, **50,** 369.
[37] H. Meier and E. Unger, *J. Radioanal. Chem.*, 1976, **32,** 413.
[38] H. Meier and E. Unger, *Staub-Reinh. Luft*, 1975, **9,** 321.
[39] K. H. Lieser, E. Breitwieser, P. Burba, M. Röber and R. Spatz, *Microchim. Acta*, 1978 **I,** 363.
[40] A. H. Pradzyński, E. Henry and J. S. Stewart, *J. Radioanal. Chem.*, 1976, **32,** 219.
[41] H. Pradzyński, R. E. Henry and J. L. S. Stewart, *Radiochem. Radioanal. Lett.*, 1975, **21,** 277.
[42] B. Holyńska, *Radiochem. Radioanal. Lett.*, 1974, **17,** 313.
[43] R. Cesareo and M. Guidotti, *Isotopenpraxis*, 1978, **14,** 353.
[44] R. Cesareo, S. Sciutti and G. E. Gigante, *Int. J. Appl. Radiat. Isot.*, 1976, **27,** 58.
[45] R. Cesareo and G. E. Gigante, *Water, Air, Soil Pollut.*, 1978, **9,** 99.
[46] B. Holyńska and J. Bisiniek, *J. Radioanal. Chem.* 1976, **31,** 159.
[47] C. Bergerioux and W. Haerdi, *Analusis*, 1980, **8,** 169.
[48] B. Holyńska, *Nucl. Techn. Groundwater Pollut. Res., Proc. Advis. Group Meet.*, 1976 179.
[49] S. M. Al-Jobobi, S. Szegedi and A. Pászit, *Radiochem. Radioanal. Lett.*, 1977, **30,** 45.
[50] M. Schiekel, K. Irmer, T. Ćechák and J. Šeda, *Isotopenpraxis*, 1979, **15,** 52.
[51] A. Markowicz, *X-Ray Spectrometery*, 1979, **8,** 14.
[52] A. Ljubicic, I. Orlic, V. Valkovic, B. Bek, S. Holjevic, N. Orlic, M. Budnar and M. Kregar, *Prog. Water Technology*, 1980, **12,** 513.
[53] S. Piórek, *Nukleonika*, 1978, **23,** 267.
[54] S. Piórek, *J. Radioanal. Chem.*, 1980, **58,** 373.
[55] S. Piórek, *Freiberg. Forschungsh.*, 1976, **C318,** 107.
[56] E. Havránek, A. Bumbálová, E. Dejmková and M. Harangozó, *Rádioaktivita a životné prostredie*, 1981, **4,** 289.
[57] E. Havránek, A. Bumbálová, M. Harangozó, E. Dejmková and F. Krečmer, *Chem. Zvesti*, 1983, **37,** 201.
[58] A. Bumbálová, E. Havránek and M. Harangozó, *Radiochem. Radioanal. Lett.*, 1982, **54,** 367.
[59] J. Bacsó, H. Kis-Varga, P. Kovács and G. Kalinka, *ATOMKI Közl.*, 1978, **20,** 237.
[60] T. Paradellis and N. Panayotakis, *J. Radioanal. Chem.*, 1980, **59,** 221.
[61] P. Nagy and A. Paszit, *Agrokém. Talajtan*, 1979, **28,** 167.
[62] A. Paszit and P. Nagy, *Izotóptechnika*, 1979, **22,** 41.
[63] G. Gomel, *Int. Symp. Anal. Chem. Explor. Min. Process. Mater.*, Pretoria, 1977, 421.
[64] N. K. Mumba, J. Csikai and G. Peto, *Radiochem. Radioanal. Lett.*, 1982, **52,** 373.
[65] L. Ahlgren and S. Mattson, *Phys. Med. Biol.* 1979, **24,** 136.

[66] L. Ahlgren, K. Liden, S. Mattson, S. Tejning, *Scand. J. Work Environ. Health*, 1976, **2,** 82.
[67] L. Ahlgren, K. Liden and S. Mattson, *9th Nord. Meet. on Clin. Phys.*, Gothenburg, Sweden, 1977.
[68] L. Ahlgren, T. Grönberg and S. Mattson, *Advances X-Ray Anal.*, 1980, **23,** 185.
[69] L. Ahlgren, B. Haeger-Aronsen, S. Mattson and A. Schütz, *Brit. J. Industrial Med.*, 1980, **37**, 109.
[70] C. D. Strehlow and Th. J. Kneip, *Am. Ind. Hyg. Assoc. J.*, 1969, **30,** 372.
[71] L. Wielopolski, D. N. Slatkin, D. Varsky, K. J. Ellis and S. M. Cohn, *IEEE Trans. Nucl. Sci.*, 1981, **NS 28,** 114.
[72] E. E. Laird, D. R. Chettle and M. C. Scott, *Nucl. Instrum. Methods Phys. Res.*, 1982, **193,** 377.
[73] J. Bacsó, *ATOMKI Közl.*, 1977, **19,** 157.
[74] R. Cesareo, *Isotopenpraxis*, 1975, **11,** 302.
[75] R. Cesareo, G. Tallarida and F. Baldoni, *Int. J. Appl. Radiat. Isotop.*, 1975, **26,** 285.
[76] Xuebeng Wang, Xinjuan Xia, Zhixiang Chen, Yaoliang Xu, Zhixuan Shi and Jianping Wu, *Huanjing Kexue*, 1981, **2,** 341.
[77] H.-K. Bothe and E. J. Lenk, *Zf I-Mitt*, 1970, **22**, Suppl., 146.
[78] H.-K. Bothe and D. Wagner, *Z. Anal. Chem.*, 1981, **306,** 15.
[79] I. Z. Vander, V. E. Gerling and V. N. Smirnov, *Zavodsk. Lab.*, 1971, 162.
[80] I. Z. Vander, V. E. Gerling, V. I. Zgardovskii, S. V. Nerpin and A. L. Tsalkin, *Geofiz. Appl.*, 1970, No. 42, 138.
[81] P. V. Kulkarni and I. L. Preiss, *J. Radioanal. Chem.*, 1975, **24,** 423.
[82] N. A. Wogman, H. G. Rieck and J. R. Kosorok, *Nucl. Instrum. Methods*, 1975, **123,** 561.
[83] R. L. Brodziński and N. A. Wogman, *Proc. Symp. on Exploration for Uranium Ore Deposits* (*Vienna* 1976), p. 447. IAEA, Vienna 1977.
[84] D. V. Yuryev and A. S. Savvatov, *Izv. Vyss. Uhebn. Zaved., Geol. Razved.*, 1980, **23,** 61.
[85] H. Hellman, *Z. Anal. Chem.*, 1971, **254,** 192.
[86] T. B. Johansson, M. Ahlber, R. Akselsson, G. Johansson and K. Malmqvist, *J. Radioanal. Chem.*, 1976, **32,** 207.
[87] V. Hnatowicz, in *Nové směry v analytické chemii II*, J. Zýka (ed.), p. 191. SNTL, Prague, 1984.
[88] S. Tanaka, Y. Hashimoto and M. Darzi, *Nucl. Instrum. Methods*, 1981, **181,** 509.
[89] M. A. Chaudri and T. Ainsworth, *Nucl. Instrum. Methods*, 1981, **181,** 333.
[90] M. E. A. Fuchs, G. Sann, J. R. Chen, R. Nobiling, B. Povh and K. Traxel, *G-I-T Labor. Med.*, 1981, **4,** 116.
[91] H. Lannefors, *Ph.D. Thesis*, Lund Univ., 1982.
[92] P. E. Artaxo Netto and C. M. Q. Orsini, *Cienc. Cult.* (*Sao Paulo*). *Suppl.*, 1982, **34,** 269.
[93] P. E. Netto Artaxo, C. M. Q. Orsini and M. H. Tabacniks, *Cienc. Cult.* (*Sao Paulo*) *Suppl.*, 1983, **35,** 545.
[94] C. M. Fou, *IEEE Trans. Nucl. Sci.*, 1981, **NS-28,** 1407.
[95] B. R. Meyer, M. Peisach and W. A. Kotze, *Nucl. Instrum. Meth., Phys. Res.*, 1982, **193,** 331.
[96] N. S. Saleh, *J. Radioanal. Chem.*, 1982, **74,** 191.
[97] M. Renan, *Radiochem. Radioanal. Lett.*, 1979, **40,** 87.
[98] V. Valković, *Analysis of Biological Material for Trace Elements Using X-Ray Spectroscopy*, CRC Press, Boca Raton, 1980.
[99] V. Valković, D. Rendić, E. K. Biegert and E. Andrade, *Environ. Int.*, 1979, **2,** 27.
[100] E. A. Kornev and A. E. Yanchukovich, *Agrokhim.*, 1980, No. 6, 134.
[101] N. F. Mangelson and M. W. Hill, *Nucl. Instrum. Methods*, 1981, **181,** 243.
[102] W. Maenhaut and L. De Reu, *IEEE Trans. Nucl. Sci.*, 1981, **NS-28,** 1386.
[103] F. Bosch, H. D. Reiss and K. Traxel, *Nucl. Sci. Appl.*, 1980, **1,** 1.
[104] S. K. Biswas, M. Abdullah, S. Akhter, S. A. Tarafdar, Khaliquzzaman and A. H. Khan, *J. Radioanal. Nucl. Chem. Articles*, 1984, **82,** 111.
[105] M. Ishikawa, K. Kitao, H. Imaseki, T. Ishii and S. Uchida, *J. Radioanal. Nucl. Chem. Articles* , 1984, **82,** 182.
[106] I. Borbély-Kiss, E. Koltay, S. László, Gy., Szabó, S. Gödény and El-Nasr S, Seif, *J. Radioanal. Nucl. Chem. Articles*, 1984, **83,** 175.
[107] G. S. Hall, N. Roach, U. Simmons, H. Cong, M. Lee and E. Cummings, *J. Radioanal. Nucl. Chem. Articles*, 1984, **82,** 329.

[108] S. Järnström, J. Dahlbacka, P. Pakarinen, V. Näntö and M. Grönroos, *J. Radioanal. Chem.*, 1983, **79**, 207.
[109] A. B. Hallak, *J. Radionanal. Chem.*, 1983, **77**, 201.
[110] R. Lecomte, S. Landsberger and S. Monaro, *J. Appl. Radiat. Isot.*, 1982, **33**, 121.
[111] I. Roelandts, P. Bartsch, A. Collignon, G. Robaye, G. Weber, J. M. Delbrouck and J. Yujie, *Radiochem. Radioanal. Lett.*, 1982, **50**, 319.
[112] N. S. Saleh, *J. Radioanal. Chem.*, 1982, **74**, 257.
[113] S. Augustyniak, L. Jarczyk, J. B. Lenszowska, E. Rokita and D. Slominska, *IEEE Trans. Nucl. Sci.*, 1981, **NS-28**, 1398.
[114] J. Tölgyessy and P. Dillinger, *Chem. Listy*, 1970, **64**, 235.
[115] J. Tölgyessy, *Jaderná energie*, 1966, **12**, 81.
[116] R. V. Dzagatspanyan, R. F. Romm and L. K. Tatochenko, *Primenenie radioaktivnykh izotopov dlya kontrolya khimicheskikh protsessov*, Goskhimizdat, Moscow, 1963.
[117] J. Tölgyessy, *Jad. Energ.*, 1970, 109.
[118] R. Jirkovský, K. Müller and J. Tomáš, *Využití radioizotopu v hutnictví a hornictví* (*Application of Radioisotopes in Metallurgy and Mining*), SNTL, Prague, 1967.

9

Sources of information

9.1 ENVIRONMENTAL CHEMISTRY — GENERAL

S. E. Jørgensen (ed.), *Application of Ecological Modelling in Environmental Management*, Elsevier, Amsterdam, 1983.

R. A. Baiki, H. M. Clark, J. P. Ferris, S. Kranse and R. L. Strong, *Chemistry of the Environment*, Academic Press, New York, 1978.

R. Baudo and H. Muntau, *Particulate Matter in the Biogeochemical Cycles of Elements*, Ellis Horwood, Chichester, 1986.

J. O'M. Brockris, *Environmental Chemistry*, Plenum Press, New York, 1977.

P. J. Craig, *The Natural Environment and the Biogeochemical Cycles*, Springer-Verlag, New York, 1980.

R. L. Swan and A. Eschenroeder (eds.) *Fate of Chemicals in the Environment: Compartmental Modelling for Predictions*, American Chemical Society, Washington, D.C., 1983.

R. A. Horne, *The Chemistry of Our Environment*, Wiley, New York, 1978.

M. Lippman and R. B. Schlessinger, *Chemical Contamination in the Human Environment*, Oxford University Press, New York, 1979.

S. E. Manahan, *Environmental Chemistry*, 4th Ed., Willard Grant, Boston, 1984.

H. W. Nürnberg and J. Vigneron (eds.), *Pollutants and their Ecotoxicological Significance*, Wiley, New York, 1985.

D. Purves, *Trace-Element Contamination of the Environment*, 2nd Ed., Elsevier, New York, 1985.

I. J. Tinsley, *Chemical Concepts in Pollutant Behaviour*, Wiley–Interscience, New York, 1979.

K. Verschneren, *Handbook of Environmental Data on Organic Chemicals*, Van Nostrand–Reinhold, New York, 1977.

9.2 ENVIRONMENTAL CHEMISTRY — WATER

J. N. Bowen, *Groundwater*, Halsted Press, New York, 1980.

J. Buffle, *Complexation Reactions in Aquatic Systems: An Analytical Approach,* Ellis Horwood, Chichester, 1987.

J. P. Riley and R. Chester (eds.), *Chemical Oceanography*, Academic Press, New York, 1978, and annual volumes.

P. R. Dugan, *Biochemical Ecology of Water Pollution*, Plenum Press, New York, 1972.

S. D. Frust and O. M. Aly, *Chemistry of Natural Waters*, Ann Arbor Science, Ann Arbor, Mich., 1981.

J. D. Hem, *Study and Interpretation of the Characteristics of Natural Water*, 2nd Ed., *U.S. Geol. Survey Paper* No. 1473; United States Geological Survey, Washington, D.C., 1970.

A. James (ed.), *An Introduction to Water Quality Modeling*, Wiley, New York, 1984.

W. Kinzelbach, *Groundwater Modeling; An Introduction with Sample Programs in Basic*, Elsevier, New York, 1986.

A. C. Sigleo and A. Hattori (eds.), *Marine and Estuarine Geochemistry*, Lewis, Chelsea, Mich., 1985.

W. Stumm and J. J. Morgan, *Aquatic Chemistry*, 2nd Ed., Wiley–Interscience, New York, 1981.

T. D. Waite, *Principles of Water Quality*, Academic Press, Orlando, Florida, 1984.

C. H. Ward, W. Giger and P. L. McCarty, *Ground Water Quality*, Wiley, New York, 1985.

9.3 ENVIRONMENTAL CHEMISTRY — AIR

A. C. Stern (ed.), *Air Pollutants*, Academic Press, Orlando, Florida, 1986.

C. De Wispelaere (ed.), *Air Pollution Modeling and Its Application*, Plenum Press, New York, 1985.

T. E. Graedel, D. T. Hawkins and L. D. Claxton, (eds.), *Atmospheric Chemical Compounds*, Academic Press, Orlando, Florida, 1986.

S. J. Eisenreich (ed.), *Atmospheric Pollutants,* Ann Arbor Science, Ann Arbor, Mich., 1981.

R. G. Barry and R. J. Chorley, *Atmosphere, Weather and Climate*, 3rd Ed., Halsted Press, New York, 1978.

B. J. Finlayson-Pitts and J. N. Pitts, *Atmospheric Chemistry: Fundamentals and Experimental Techniques*, Wiley–Interscience, New York, 1986.

T. Godish, *Air Quality*, Lewis, Chelsea, Mich., 1985.

A. H. Legge and S. V. Kruppa, *Air Pollutants and their Effects on the Terrestrial Ecosystem*, Wiley, New York, 1986.

A. R. Meetham, D. W. Bottom, S. Clayton, A. Henderson-Sellers and D. Chambers, *Atmospheric Pollution*, Pergamon Press, Elmsford, N.J., 1981.

E. Mesyaros, *Atmospheric Chemistry*, Elsevier–North Holland, New York, 1981.

J. H. Seinfeld, *Atmospheric Chemistry and Physics of Air Pollution*, Wiley, New York, 1985.

A. C. Stern, R. W. Baubel, D. B. Turner and D. L. Fox, *Fundamentals of Air Pollution*, Academic Press, Orlando, Florida, 1984.

P. J. Lioy and J. M. Daisey (eds.), *Toxic Air Pollution*, Lewis, Chelsea, Mich., 1986.

K. Work and C. F. Warner, *Air Pollution: Its Origin and Control*, 2nd Ed., Harper and Row, New York, 1981.

9.4 ENVIRONMENTAL CHEMISTRY — SOIL

R. H. Dowdy, *Chemistry in the Soil Environment*, American Society of Agronomy, Soil Science Society of America, Madison, Wisc., 1981.

A. C. Fortescue, *Environmental Geochemistry*, Springer-Verlag, New York, 1980.

G. H. Bolt (ed.), *Soil Chemistry*, Elsevier–North Holland, New York, 1979.

G. H. Speidel and A. F. Angew, *The Natural Geochemistry of Our Environment*, Westview, Boulder, Colo., 1982.

F. J. Stevenson, *Humus Chemistry: Genesis, Composition, Reactions*, Wiley–Interscience, New York, 1982.

9.5 ANALYTICAL TECHNIQUES — GENERAL

J. Albaiges, R. W. Frei and E. Merian, *Chemistry and Analysis of Hydrocarbons in the Environment*, Gordon and Breach, New York, 1983.

W. R. Loing (ed.), *Analytical Chemistry Instrumentation*, Lewis, Chelsea, Mich., 1986.

J. Albaiges (ed.), *Analytical Techniques in Environmental Chemistry*, Pergamon Press, Elmsford, N. J., 1982.

Annual Book of ASTM Standards, Water and Atmospheric Analysis, American Society for Testing and Meterials, Philadelphia, annually.

D. L. Tsalev, *Atomic Absorption Spectrometry in Occupational and Environmental Health Practice*, CRC Press, Boca Raton, Florida, 1984.

E. P. Bertin, *Introduction to X-ray Spectrometric Analysis*, Plenum Press, New York, 1978.

K. Beyermann, *Organic Trace Analysis*, Ellis Horwood, Chichester, 1984.

G. E. Schweitzer and J. A. Santolucito (eds.), *Environmental Sampling for Hazardous Materials*, American Chemical Society, Washington, D.C., 1984.

G. W. Ewing, *Instrumental Methods of Chemical Analysis*, 5th Ed., McGraw-Hill, New York, 1985.

S. S. M. Hassan, *Organic Analysis Using Atomic Absorption Spectrometry*, Ellis Horwood, Chichester, 1984.

E. Jungreis, *Spot Test Analysis; Environmental, Forensic and Geochemical Applications*, Wiley, New York, 1985.

C. Liteanu and I. Rîcă, *Statistical Theory and Methodology of Trace Analysis*, Ellis Horwood, Chichester, 1980.

Z. Marczenko, *Separation and Spectrophotometric Determination of Elements*, Ellis Horwood, Chichester, 1986.

F. W. Karasek, O. Hutzinger and S. Safe (eds.), *Mass Spectrometry in Environmental Sciences*, Plenum Press, New York, 1984.

Measurement, Detection and Control of Environmental Pollutants, IAEA-SM-20516, International Atomic Energy Agency, Vienna, 1976.
B. S. Middleditch, S. R. Missler and H. B. Hines, *Mass Spectrometry of Priority Pollutants*, Plenum Press, New York, 1981.
J. C. Miller and J. N. Miller, *Statistics for Analytical Chemistry*, Ellis Horwood, Chichester, 1984.
J. Minczewski, J. Chwastowska and R. Dybczyński, *Separation and Preconcentration Methods in Inorganic Trace Analysis*, Ellis Horwood, Chichester, 1982.
J. R. Moody, *Sampling and Sample Storage for Trace Elemental Analysis*, Ellis Horwood, Chichester, 1985.
J. R. Rhodes, in *Practical Aspects of Energy-Dispersive X-Ray Emission Spectrometry*, Report IAEA-216, International Atomic Energy Agency, Vienna, 1979.
J. C. Van Loon (ed.), *Selected Methods of Trace Metal Analysis: Biological and Environmental Samples*, Wiley, New York, 1985.
J. F. Lawrence (ed.), *Trace Analysis*, Vol. 4, Academic Press, New York, 1985.
G. D. Christian, Jr. and J. B. Callis (eds.), *Trace Analysis: Spectroscopic Methods for Molecules*, Wiley–Interscience, New York, 1986.
D. A. Kurtz (ed.), *Trace Residue Analysis, Chemometric Estimation of Sampling, Amount and Error*, American Chemical Society, Washington, D.C., 1985.
A. Varma, *Handbook of Atomic Absorption Analysis*, CRC Press: Boca Raton, Florida, 1984.
J. Veselý, D. Weiss and K. Štulík, *Analysis with Ion-Selective Electrodes*, Ellis Horwood, Chichester, 1978.
T. G. Dzubay (ed.), *X-Ray Fluorescence Analysis of Environmental Samples*, Ann Arbor Science, Ann Arbor, Mich., 1977.

9.6 ANALYTICAL TECHNIQUES — WATER

T. R. Compton, *Determination of Organic Substances in Water*, New York, 1985.
R. W. Frei and U. A. Brinkman, *Analysis and Chemistry of Water Pollutants*, Gordon and Breach, New York, 1983.
J. Krajca, *Natural Water Sampling*, Ellis Horwood, Chichester, 1986.
Methods for Chemical Analysis of Water and Wastes, Environmental Monitoring and Support Laboratory, Office of Research and Development, U.S. Environmental Protection Agency, Cincinnati, Ohio, 1979 and addendum.
Standard Methods for the Examination of Water and Wastewater, 16th ed., American Public Health Association, American Water Works Asssociation, Water Pollution Control Federation, Washington, D.C., 1985.
G. C. Leppard, (ed.), *Trace Element Speciation in Surface Waters*, Plenum Press, New York, 1983.

J. C. Van Loon, *Chemical Analysis of Inorganic Constituents of Water*, CRC Press, Boca Raton, Florida, 1982.
R. A. Minear and L. H. Keith, (eds.), *Water Analysis*, Academic Press, Orlando, Florida, 1984.
R. K. Winge, J. M. Katzenberger and V. A. Fassel, *Development and Application of an Inductively-Coupled Plasma Analytical System to the Simultaneous Multielement Determination of Trace Elemental Pollutants in Water*, Report EPA-IAG-D6-0417, U. S. Environmental Protection Agency, Washington, D.C., 1977.

9.7 ANALYTICAL TECHNIQUES — AIR

F. Schneider, H. W. de Doning and L. J. Brasser (eds.), *Air Pollution: Reference Measurement Methods and Systems*, Elsevier–North Holland, New York, 1978.
R. M. Harrison and R. Perry (eds.), *Handbook of Air Pollution Analysis*, Chapman & Hall, New York, 1986.
Instrumentation for Environmental Monitoring–Air, Lawrence Berkely Laboratory, University of California, Berkely, 1986 and addendum.
W. Leithe, *The Analysis of Air Pollutants*, Ann Arbor Science, Ann Arbor, Mich., 1971.
E. J. M. Katz (ed.), *Methods of Air Sampling and Analysis*, American Public Health Association, Washington, D.C., 1977.
K. Oikawa, K. *Trace Analysis of Atmospheric Samples*, Halsted Press, Tokyo, 1977.
R. Perry and R. J. Young, *Handbook of Air Pollution Analysis*, Wiley, New York, 1977.
A. C. Stern, *Air Pollution, Vol. III, Measuring, Monitoring and Surveillance of Air Pollution*, 3rd Ed., Academic Press, New York, 1976.

9.8 ANALYTICAL TECHNIQUES — SOIL

K. H. Head, *Manual of Soil Laboratory Testing*, Vols. I, II, III, Wiley–Halsted, New York, Vol. III, 1986.
A. L. Page, R. H. Miller and D. R. Keeney, *Methods of Soil Analysis*, American Society of Agronomy, Soil Science Society of America, Madison, Wisc., 1983.
K. A. Smith, *Soil Analysis: Modern Instrumental Techniques*, Dekker, New York, 1983.

9.9 ANALYTICAL TECHNIQUES — NUCLEAR

R. H. Filby (ed.), *Atomic and Nuclear Methods for Fuel and Energy Research*, Plenum Press, New York, 1982.
V. Balek and J. Tölgyessy, *Emanation Thermal Analysis and Other Radiometric Emanation Methods*, Elsevier, Amserdam, 1984.

T. Braun and J. Tölgyessy, *Radiometric Titrations*, Pergamon Press, Oxford, 1967.

G. R. Choppin and J. Rydberg, *Nuclear Chemistry*, Pergamon Press, Oxford, 1980.

H. A. Das, A. Faanhof and H. van der Sloot, *Environmental Radioanalysis*, Elsevier, Amsterdam, 1983.

G. G. Eichholz, *Environmental Aspects of Nuclear Power*, Lewis, Chelsea, Mich., 1985.

A. Brodsky (ed.), *Handbook of Radiation Measurement and Protection*, CRC Press, Boca Raton, Florida, 1979.

R. L. Kathren, *Radioactivity in the Environment*, Harwood, New York, 1984.

G. F. Knoll, *Radiation Detection and Measurement*, Wiley, New York, 1979.

P. Kruger, *Principles of Activation Analysis*, Wiley–Interscience, New York, 1971.

M. F. L'Annunziata, *Radionuclide Tracers; Their Detection and Measurement*, Academic Press, Orlando, Florida, 1986.

A. Martin and S. A. Harbison, *An Introduction to Radiation Protection*, Chapman & Hall, New York, 1986.

V. Majer, *Grundlagen der Kernchemie*, Barth, Leipzig, 1982.

Nuclear Techniques in Environmental Pollution, International Atomic Energy Agency, Vienna, 1971.

Nuclear Methods in Environmental Research, CONF. 771072, NTIS, U.S. Dept. of Commerce, Springfield, VA, 1977.

G. G. Eichholz and J. W. Poston (eds.), *Principles of Nuclear Radiation Detection*, Lewis, Chelsea, Mich., 1985.

Proceedings of the Third International Conference on PIXE and its Analytical Application, Heidelberg, July, 1983; Nucl. Instrum. Methods Phys. Rev., 1984.

E. A. Evans and M. Miamatsu, (eds.), *Radiotracer Techniques and Applications*, Dekker, New York, 1977.

J. R. Rhodes, J. A. Stout, J. G. Schindler and S. Piorek, *Portable X-Ray Survey Meter for in situ Trace Element Monitoring of Air Particulates*, American Society for Testing Materials, Philadelphia, 1983.

S. M. Roa, A. B. Majali, R. G. Desphande and T. S. Murty, *Industrial Applications of Radioisotopes and Radiation*, Wiley–Halsted, New York, 1986.

J. Růžička and J. Starý, *Substoichiometry in Radiochemical Analysis*, Pergamon Press, Oxford, 1968.

L. Salmon, *Instrumental Neutron Activation Analysis in Environmental Studies of Trace Element*, AERE-R-7859, Atomic Energy Research Establishment, Harwell, 1975.

V. Schultz and F. W. Wicker, *Radioecological Techniques*, Plenum Press, New York, 1982.

P. Sims and F. A. Rickey, *The Multielement Analysis of Drinking Water Using Proton-Induced X-Ray Emission (PIXE)*, Environmental Protection Agency, U.S. Dept. of Commerce, Springfield, VA, 1978.

J. Tölgyessy, T. Braun and M. Kyrš, *Isotope Dilution Analysis*, Pergamon Press, Oxford, 1972.

J. Tölgyessy and M. Kyrš, *Radioanalytical Chemistry*, Ellis Horwood, Chichester, in the press.

J. Tölgyessy and S. Varga, *Nuclear Analytical Chemistry*, Vols. I–V; University Park, Baltimore; Publishing House Slovak Academy of Science, Bratislava, 1976.

J. Tölgyessy, J. V. Zakovlev and G. N. Bilimovich, *Diagnostika okruzkayuskei sredy radioanaliticheskimi metodami* (Diagnosis of the Environment with Radioanalytical Methods), Energoatomizdad, Moscow, 1985.

N. Tsoulfanidis, *Measurement and Detection of Radiation*, McGraw-Hill, New York, 1983.

C. H. Wang, D. L. Willis and W. D. Loveland, *Radiotracer Methodology in the Biological, Environmental and Physical Sciences*, Prentice-Hall, Englewood Cliffs, N.J., 1975.

Index